Reclaiming Our Planet

Reclaiming Our Planet

How Environmental History Can Help Solve the Climate Crisis

Alexander Gates

ROWMAN & LITTLEFIELD
Lanham • Boulder • New York • London

Published by Rowman & Littlefield
An imprint of The Rowman & Littlefield Publishing Group, Inc.
4501 Forbes Boulevard, Suite 200, Lanham, Maryland 20706
www.rowman.com

86-90 Paul Street, London EC2A 4NE

British Library Cataloguing in Publication Information Available

Library of Congress Cataloging-in-Publication Data

Names: Gates, Alexander E., 1957- author.
  Title: Reclaiming our planet : how environmental history can help
    solve the climate crisis / Alexander Gates.
  Description: Lanham : Rowman & Littlefield, [2024] | Includes
    bibliographical references and index. | Summary: "The climate crisis has
    been portrayed as deadly to humans, and it could be. However, serious
    environmental and resulting public health crises have run rampant for a
    century or more. This book looks at how several of these crises
    developed and were miraculously resolved and then compares how the
    climate crisis is being addressed"-- Provided by publisher.
  Identifiers: LCCN 2023058576 (print) | LCCN 2023058577 (ebook) | ISBN
    9781538179673 (cloth) | ISBN 9781538179680 (epub)
  Subjects: LCSH: Human ecology--History. | Climate change mitigation.
  Classification: LCC GF13 .G38 2024 (print) | LCC GF13 (ebook) | DDC
    304.2/8--dc23/eng/20240418
  LC record available at https://lccn.loc.gov/2023058576
  LC ebook record available at https://lccn.loc.gov/2023058577

♾️™ The paper used in this publication meets the minimum requirements of American
National Standard for Information Sciences—Permanence of Paper for Printed Library
Materials, ANSI/NISO Z39.48-1992.

This book is dedicated to Dr. Jill C. Stein

Contents

Acknowledgments

This book benefited greatly from reviews of early drafts of several chapters by Dr. Cy Stein. Sections of other chapters were extensively reviewed by Dr. Colin Gates and Dr. Peter Kowal to improve their accuracy and ensure the latest information was included. Discussions with these people and Dr. David Valentino helped crystallize many of the ideas. Their input and feedback are appreciated. Dr. Colin Gates assisted with figures and references.

Chapter 1

The Reality of Climate Change

CLIMATE CHANGE VIEWS AND WHY
THIS BOOK IS DIFFERENT

From the beginning of the species, humans have been destructive to the natural environment. In the beginning, the destruction was simple, like uprooting trees to make tools or clearing land or using fire for cooking and heat or even trampling vegetation in an encampment. This destructiveness has intensified to the devastation of emitting toxic chemicals and radiation to the environment and facilitating the introduction of dangerous invasive species. When there were fewer people on the planet, the process of destruction by humans and subsequent recovery was sustainable, with natural processes removing the damage relatively quickly. It is estimated that the human population ranged between 2 and 5 million from 50,000 to 10,000 years ago, which allowed the damage to be recoverable by natural processes even if it was substantial.[1] With the population currently approaching 8 billion people, regardless of the validity of these population estimates, the earth is well in excess of a sustainable human population. Further, humans are emitting much more dangerous pollutants than we did in the distant past. At the current unsustainable population, even small emissions of not-so-dangerous pollutants in transportation or yardwork or household garbage by such enormous numbers of people still amounts to overwhelmingly excessive pollution that can have devastating effects on the natural environment. This overwhelming quantity of release of apparently benign substances is what is causing the climate crisis humans now face.

Although the science behind a potential climate crisis has been known since the early 1950s when it was named the "industrial effect" it was generally not considered a danger.[2] The earth's climate is so voluminous and so readily exchanges chemicals, like CO_2, with the oceans, it was generally regarded that human input was too small to make a difference. It was only Roger Revelle, the acknowledged grandfather of climate change, who sounded the warning of excessive use of fossil fuels as early as the late 1960s. All the way to about 1980, it seemed that the conventional wisdom that human input was insignificant was correct. But then the climate began to show the damage being done to it and the concern has ballooned. Since then, climate change has developed into the environmental crisis of our age if not the most concerning human crisis ever. Former vice president Al Gore won the Nobel Prize for just bringing climate change to public consciousness through an award-winning documentary in addition to books and public appearances.[3] More recently, young Swedish environmental activist Greta Thunberg gained worldwide attention for speeches and rallies against climate change. Governments have also taken steps to address this climate crisis. Most industrialized nations have now passed or are passing legislation to address climate change and many have already taken action to reduce the use of fossil fuels. The United Nations has held several conventions and developed protocols and agreements to address climate change. This climate battle has so corralled public attention that virtually all other environmental issues, regardless of how dire, have faded into the background.

It is a welcome development that the public regards an environmental issue as this important. A problem for the public in dealing with climate change is that there is so much misinformation and even disinformation in the available public media and social media that it has led to significant confusion on a national to global scale. In addition to the politically charged rhetoric, there are numerous books and other publications by famous people like Bill Gates and many others who may not have the scientific background in climate science to provide a clear and accurate view of the issue and all of its complexities. Many scientists release technical studies on climate change that are not easily digested by the general public, possibly leading to even more confusion. Other scientists may allow personal preferences into their published analyses that are also misleading. There are so many competing factors in the complex relationships in climate change that it is difficult to make accurate predictions even if all research is done and reported in the most accurate and ethical manner. Finally, evaluations and predictions are done in a vacuum because humans have never faced such a dire situation at such a large scale before.

As a result of all of these competing scientific factors, widespread misinformation and disinformation, public opinions on climate change cross the entire range of possibilities. Some people have accepted that climate change is a hoax and was invented for some nefarious reason. These people are called climate change deniers. Some people accept that climate change is occurring but believe that it is natural and is not the result of human activity. They argue that the climate changes all of the time with or without human influence. Further, in this view, it is better to have a warmer earth than a colder one. Some people even regard climate change as beneficial because the resulting melting of the Arctic ice sheet will open new shipping routes that connect the Atlantic and Pacific Oceans. They also argue that the poleward shifting of climate belts might make more areas habitable and may even make it possible to produce more food with larger areas to farm. Many other people believe that climate change is a danger but don't do anything about it and expect the government to address the problem on their own. An increasing number of people think that climate change can only be slowed down but that it will continue unabated, basically forever. Finally, there is a group who think that there is no way to address climate change and that we are all doomed.

This book is designed to help people from all of these viewpoints to understand the severity of the climate crisis, its potential impact on people and the environment, and the steps that are being taken and can be taken to address it. However, rather than thinking in a vacuum, this book examines other current and previous environmental crises that humans have faced and defeated, bringing them to a safe resolution. These situations are presented as case studies describing the development and severity of the problem and the subsequent impressive and directed efforts by groups and individuals that resulted in or are resulting in a full resolution of the very serious environmental problems. They include how we, as the public, forced politicians and industry to resolve these problems in the quickest and most effective manner regardless of the cost and ideological resistance. It is these victories that can give all people the confidence to believe that the climate crisis can be resolved without causing a massive extinction event that includes the human race. The case studies can also serve as guides by which concerned individuals can enact fundamental changes to the environment through education, physical action, personal responsibility, and public awareness. This book deals with climate change as just one of a number of environmental problems that humans have faced and are facing.

IS CLIMATE CHANGE REAL?

The science behind our current situation with climate change involves the ever-increasing concentration of carbon dioxide (CO_2) in the atmosphere. CO_2 is a greenhouse gas, and as such, it allows the passage of sunlight through the atmosphere and to the earth's surface.[4] The incoming sunlight heats the surface and the heat is released to the atmosphere but is trapped by the CO_2. The atmosphere consequently slowly heats up. This situation is analogous to a greenhouse where the sunlight can pass through the glass roof and walls into the house but the heat inside cannot escape through them. For this reason, CO_2 is termed a greenhouse gas. There are much more potent greenhouse gases like methane, which is 23 times as powerful at trapping heat, but most are in such low concentrations compared to CO_2 that their impact is considered negligible. The problem is that the CO_2 that was removed from the atmosphere by biologic activity and sequestered within the earth through natural processes over many tens to hundreds of millions of years is now being released back to the atmosphere through the burning of fossil fuels at an alarming rate. This essentially reverses the natural evolution of the atmosphere and life on earth.

The earth has an envelope of gases that are held to the surface by gravity that is called the atmosphere.[5] With the heat from the sun, the atmosphere strongly modulates the temperature on the planet. The earth's atmosphere is composed of 78 percent nitrogen, 21 percent oxygen, and 1 percent of all other gases combined including only about 417 parts per million (ppm) of CO_2. The average temperature at the earth's surface is a comfortable 59°F (15°C). Without an atmosphere, the earth would have similar temperatures to the moon as it is a similar distance from the sun.[6] The temperatures at the equator of the moon reach a high of a scorching 250°F (120°C), and a low of an absurdly cold -208°F (-130°C) at the surface. Away from the equator, it is extremely cold all of the time with temperatures reaching as low as -410°F (-250°C) at the poles. In contrast, Venus is a bit closer to the sun than the earth but has an atmosphere composed of 96.5 percent CO_2 and only 3.5 percent nitrogen, basically all CO_2.[7] As a result of global warming from this extreme amount of greenhouse gas, the surface temperature on Venus ranges from about 820 to 900°F (438–482°C) with an average of a scorching 847°F (453°C). This is hot enough to melt lead. These examples show the importance of greenhouse gases in regulating surface temperatures at similar distances from the sun.

The response to the question of whether there is currently climate change on Earth is a definite yes. Over the past 100 years, the average temperature of the earth has increased by at least 2°F (1.1°C).[8] The beginning part of this change has to be estimated because worldwide reliable data collection arrays were not available at that time. However, it can be well documented that the global temperature has greatly accelerated since about 1980 and has risen by about 1°F (0.55°C) since 1991 alone. The concern is that this change in temperature is dramatic and certainly not within the scope of normal variations.

These data also somewhat address whether the current change is the result of natural variation. The current change is much greater than normal climate variation over the past 2,000 years. Further, the change is in lockstep with increasing CO_2 levels.[9] CO_2 in the atmosphere was about 280 ppm in 1800, 295 ppm in 1900, about 315 ppm in 1950, and then skyrocketed to 417 ppm

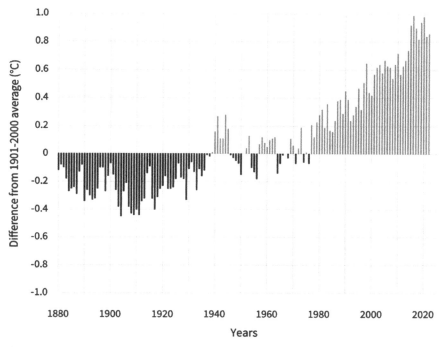

Figure 1.1. **Bar graph of average annual temperatures from 1880 to 2022 with 0 at the average temperatures from 1901 to 2000. Light gray bars extending up from the average are above-average years and black bars extending down are below-average years. Note the rapid increase in temperature after 1980.**
Source: NOAA.

or so today. The most rapid increase in CO_2 levels has been since 1980. In fact, based on analysis of bubbles in ice cores, over the past 800,000 years, the highest atmospheric CO_2 content was about 300 ppm prior to the twentieth century.[10]

Further, the increasing CO_2 is definitely from the burning of fossil fuels and not from the natural respiration of plants and bacteria. Scientists can determine the age of ancient objects using carbon-14 (C14) isotopic dating.[11] This is because carbon from natural respiration of live sources has a certain amount of C14 which is radioactive and unstable slowly decaying to nitrogen-14 (N14). There is no C14 in fossil fuels because they are so old that it has all decayed away to N14 long ago. As such, they are considered "dead sources." This science was discovered by Dr. Hans Suess in the early 1950s. The ratio of C14 in CO_2 measured anywhere on the earth has been steadily decreasing at the same rate as the atmospheric increase in CO_2. This proves

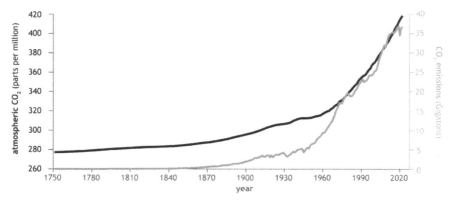

Figure 1.2.   Graph of global CO$_2$ emissions and atmospheric content from 1751 to 2022. The black line is atmospheric CO$_2$ and uses the left axis scale and the gray line is anthropogenic CO$_2$ emissions and uses the right axis scale.
Source: NOAA.

that the increased CO$_2$ in the atmosphere is certainly anthropogenic and from the burning of fossil fuels, not from natural sources.

Even if this is natural climate change, as some people propose, natural climate change is not always a desirable development. For example, the earth entered a natural cooling period at about 1100 AD that became pronounced from about 1400 to 1850–1900. This period has been termed "The Little Ice Age." The cold temperatures during this period created great hardships for northern settlements and resulted in many human deaths.[12] As the climate warmed up, the Western Hemisphere warmed quicker than the Far East. This perturbed the global weather patterns and caused a very strong El Niño condition. As a result, central China experienced an extreme drought from 1876–1878 that devastated the residents.[13] There is documentation of some towns not receiving a single drop of rain for as long as 340 days. This drought reduced the availability of food causing a devastating famine. This normally lush area had not a blade of grass for as far as the eye could see and the bark was stripped from all of the trees, all eaten by starving people. There are cases of parents selling their children for food. Some people ate dirt which kept them alive temporarily before it killed them. Whole villages are reported to have committed suicide rather than starve to death. Cannibalism became rampant with roving gangs attacking, killing, and eating people. There are even stories of starving people digging up recently buried corpses and eating them. In all, as many as 13 million people perished and globally, it was between 30 and 60 million or about 3 percent of the earth's population. The damage to plants and animals is incalculable.

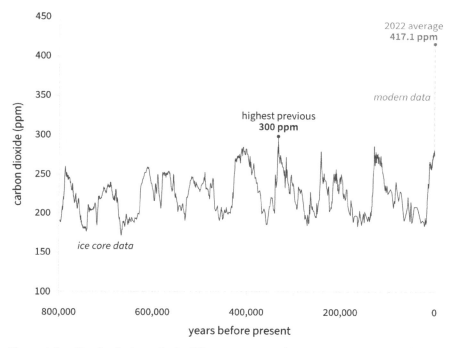

Figure 1.3. **Graph of atmospheric CO₂ content over the past 800,000 years using air bubbles trapped in glacial ice cores (solid line) and modern atmospheric measurements (dashed lines).**
Source: NOAA.

Of course, modern global trade practices should prevent such calamities even if the climate changes radically on the local or regional scale. However, in the more distant past, the earth has been plagued with natural and regular ice ages which are extreme forms of natural climate change. From about 22,000 years ago to about 12,000 years ago, the earth experienced its latest ice age.[14] Continental ice sheets up to 1 mile (1.6 km) thick extended from the North Pole to New York City and across all of northern Europe and Asia and marginally inhabitable tundra conditions extended hundreds of miles to the south of them. Truly inhabitable conditions were compressed to an area a little beyond the earth's tropics, at most. Only fauna and flora that could quickly migrate long distances survived this devastation. This was not the only ice age. In fact, the earth is currently considered to be in an interglacial period with a new ice age potentially developing at any time. Imagine the devastation an ice age would inflict on modern society.

In a worst-case scenario, natural climate change can be even more devastating than an ice age or the current climate crisis. The worst extinction event ever in the history of the planet was caused by natural global warming. During the great Permian extinction about 250 million years ago, about 96 percent of marine life and about 75 percent of terrestrial life went extinct.[15] At that time, all of the landmasses were conglomerated into one supercontinent called Pangea. With only one continent, there could be only one ocean, Panthalassa, which circulated very sluggishly. This continental geometry was stable for a while but a huge volcanic province developed in western Siberia that emitted enormous amounts of CO_2 and other chemicals to the atmosphere. The chemistry of the atmosphere progressively changed, causing global warming which was accelerated by catastrophic release of methane from the seafloor. It is estimated that the average temperature of the earth increased by about 8°F (4.5°C). This climate change took about 40,000 years. The current rate of climate change is much faster.

The message from these natural examples is that natural climate change, like many other natural disasters, is not always good. However, by considering the current climate crisis from another point of view, there appears to be some valuable information to be gained. If the current climate change is being caused by human activity, then humans have the ability to adjust the global climate as needed once the current situation is controlled. If a new ice age initiates in the future, humans will be able to adjust the amount of greenhouse gas in the atmosphere by the amount of fossil fuel consumed to prevent or control it. Right now, the earth is getting too warm, so steps must be taken to halt the trend and then cool it to an acceptable level. In this case, the past 100 years of global warming might be viewed as an experiment rather than a catastrophe. The results of this experiment can be used to better care for the earth and humankind in the future.

IS CLIMATE CHANGE A BAD THING?

Are the effects of climate change beneficial as some have proposed? The answer is that perhaps the one or two examples touted by the proponents might be somewhat beneficial but most of the results are not good, at minimum, and if they continue, they will be catastrophic. The melting of the polar ice caps as a result of the increasing temperatures and consequent flooding of coastal communities is widely recognized and well underway. Global sea level has risen about 9 inches (23 cm) since 1930 but the rate of increase has greatly accelerated since about 1980.[16] Part of the melting of the ice caps and alpine glaciers is also the result of excessive urban and industrial soot production. The black soot changes the reflectivity (or surface albedo) of the ice it lands on, causing more surface heating than would normally be expected if the snow and ice remained white and reflected back the incoming sunlight.

The exact amount of time it will take for the earth's glacial ice to melt is constantly being revised, with some claiming it will be gone within this century. The cost of the resulting coastal flooding to the world economy as the sea level rises is also a matter of speculation but has been estimated as high as $100+ trillion (US).[17] There will be exceptional efforts to prevent the flooding, loss of infrastructure, and relocation of major parts of some of the largest cities in the world. This will be devastating to coastal communities but why should people who live inland worry about climate change?

Another impact of climate change is the increase in temperature of the ocean surface.[18] Since 1910, the average surface temperature of the oceans has increased by about 2°F (1.1°C), similar to the air temperature. The increase was not dramatic until about 1980 but it has accelerated ever since. The danger of this temperature increase is that tropical weather systems including hurricanes, cyclones, and typhoons strengthen if the ocean surface temperature is greater than 79°F (26°C) but weaken if the temperature is less even if all other weather conditions favor strengthening. With ocean surface warming, the area favoring strengthening of large storms is expanding. This means that more powerful storms will persist farther from the tropics and impact areas that were previously generally spared, such as the northeastern United States. Higher temperatures also mean that large storms will be able

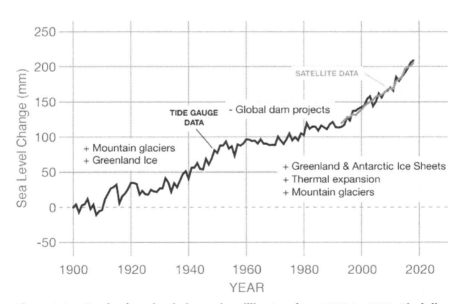

Figure 1.4. Graph of sea level change in millimeters from 1900 to 2019. Black line shows measurements from tidal gauges and the gray line is from satellite data. Note accelerated rise rate after 1980, doubling the total rise in 39 years.
Source: NOAA.

to increase in intensity much quicker than usual, a phenomenon called rapid intensification. For example, in 2017, the peak wind speeds of Hurricane Maria increased from 85 miles per hour (140 kph) (Category 1) to 165 miles per hour (270 kph) (Category 5) in just twenty-four hours right before making landfall in Puerto Rico.[19] In 2022, Hurricane Ian intensified from a tropical storm with 45 mile per hour (72 kph) winds on Sunday, September 25 to a Category 3 hurricane by Tuesday, September 27 when it struck Cuba.[20] During this time, it experienced a 67 percent strengthening in less than 22 hours. The winds slowed to 115–120 miles per hour (185–193 kph) over Cuba only to re-intensify to 155 miles per hour (250 kph) by the next day (September 28) when it made landfall in Florida, a second rapid intensification.

There has also been a general increase in the number of tropical cyclones, including Atlantic hurricanes especially since 1996.[21] However, cyclone development depends on many factors, allowing numbers to vary wildly from year to year. That is why it has been difficult to directly correlate them to progressive climate change. Just considering the Philippines, however, provides a much more convincing account.[22] The Philippines lie directly in the path of most western Pacific tropical storms and typhoons so it is a good monitor of storm activity. In the 1950s and 1960s, the Philippines experienced 18 and 21 storms per decade, respectively, with 2,362 and 4,822 storm-related deaths per decade, respectively. Since then, tropical storms and typhoons striking the Philippines have been two to three times as common every decade, reaching 91 in the 2010s and causing 12,376 deaths. Similarly, new records for the number of Atlantic hurricanes are set every few years, and they are much more prevalent than in the past. More coastal storms result in more coastal damage costing many billions of dollars and causing many more deaths and injuries. However, the larger hurricanes can also penetrate deeper inland and inflict damage there. This means that more than just coastal cities are and will be suffering from climate change impacts.

The intensity and direction of deeper ocean currents are also impacted by global warming. There are lateral currents like the Gulf Stream as well as vertical currents marked by areas in the oceans with upwelling of deep, cold ocean waters. These areas of upwelling contain nutrient-rich waters and, as a result, have high productivity of marine life. There are also areas of descending or sinking surface waters. These ocean currents can weaken, strengthen, and shift locations as a result of climate change, and they significantly impact weather and climate patterns locally and globally. El Niño and La Niña in the Pacific Ocean are infamous conditions of shifted ocean currents and they cause worldwide changes in temperatures, precipitation, and weather patterns. These two conditions reflect shifting of Pacific Ocean circulation from normal upwelling of deep ocean water off South America to abnormal (El Niño) upwelling near Australia. Over the past seventy years, during the

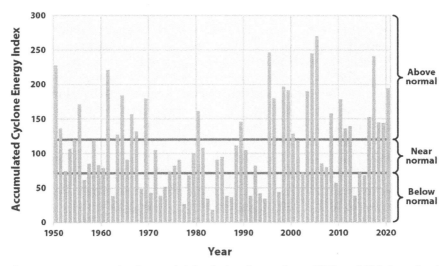

Figure 1.5. Bar graph of annual Atlantic hurricanes from 1950 to 2021 (gray bars) showing normal, above-average, and below-average years. Note the significant increase in annual hurricanes after 1995.
Source: NOAA.

rapid increase in air and ocean surface temperatures, there have also been changes in these oscillations.[23] The La Niña conditions dominated in the 1950s through 1970s and were much more common, stronger, and longer in duration than they are now. In contrast, El Niño conditions are now much more common and much stronger. The past few events have caused significant worldwide disruption of climate patterns resulting in severe droughts in some areas and catastrophic flooding in others. They impact weather across the continents, not just in coastal communities.

The warmer climates are shifting rainfall patterns and enhancing evaporation which increase the likelihood and intensity of wildfires. Wildfires were quite problematic when controlled fires were used to clear land in historical times and commonly got out of control, wreaking havoc and causing extensive death and destruction. However, firefighting efforts have improved over the years, especially with the addition of aircraft-delivered water and fire retardants. As a result, the number and extent of wildfires in many areas including the United States decreased into the 1950s and remained low through the 1970s. However, despite even more effective techniques, coordination of efforts and better fire extinguishing chemicals, in the early 1980s the number and extent of wildfires began to increase steadily and dramatically.[24] The average acreage of individual wildfires in the United States increased fourfold between 1983 and 2022, and many years the increase was fivefold.

The total number of wildfires in the western United States tripled over that period.[25] In the spring of 2023, wildfires in Quebec, Canada, produced massive amounts of smoke that was swept into the northeastern United States producing unhealthy and apocalyptic-looking conditions across the area. Unlike older wildfires that destroyed communities and caused many human deaths, the recent wildfires have mainly destroyed woodlands. However, if the increase in wildfires continues, they may begin to encroach on populated areas in the United States and claim more human lives.

There have been an increasing number of enormous wildfires around the world over the past several decades as well.[26] The 1987 Daxing'anling Wildfire or Great Black Dragon Fire struck northeast China and part of Siberia from May 6 to June 2. It killed more than 200 people, injured at least 250 and burned in excess of 32.1–37 million acres (13–15 million ha) of woodland. The major 2003 Siberian Taiga Fires burned 47 million acres (19 million ha) of woodland. From February 7 to March 14, 2009, at least 1.1 million acres (0.45 million ha) of Australian bush were burned, destroying 3,500 buildings and other structures and killing 173 people. At least 8.4 million acres (3.4 million ha) of woodland was burned in the 2014 Northwest Territories of Canada Fires. The Amazon fires burn enormous areas of the Brazilian jungle every year but in 2019, more than 40,000 fires burned 2,240,000 acres (906,000 ha), which was the highest total in more than a decade. These major wildfires were dwarfed by the 2019–2020 Australian Bush fires, which burned some 60 to

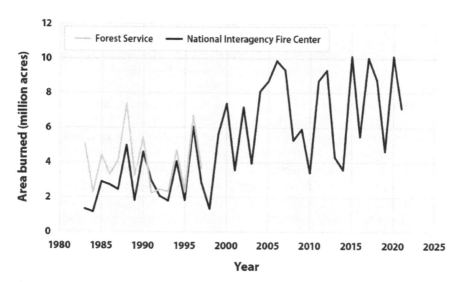

Figure 1.6.   Graph of annual area burned by wildfires in the United States from 1983 to 2022 using the two sources noted.
Source: EPA.

84 million acres (24.3 to 33.8 million ha) and killed 479 people. Wildfires are dangerous to all people across all continents, produce excessive smoke that is unhealthy to breathe, cost a tremendous amount in lost resources, and are expensive and time-consuming to extinguish. People well away from coastal areas are negatively impacted by the increasing number and size of wildfires. Wildlife is even more profoundly impacted.

Every year new and unanticipated societal problems arise as the result of climate change in addition to the known problems which continue to worsen. The summer of 2022 was very hot and dry, producing new record high temperatures. For example, the temperature reached 104.1°F (40.06°C) in London, UK, the highest temperature ever recorded in that city. Later in the summer, it was so hot in California that the electrical grid could not keep up with demand, and rolling blackouts ensued. Perhaps the most concerning of the new trends was the drying and lowering of the water level of major rivers worldwide. In the United States, as the result of a lack of rain and months of hot, dry weather across the Midwest and northern Great Plains, the Mississippi River dried to historic low levels.[27] More than 60 percent of the Midwest and northern Great Plains experienced drought conditions with below-average rainfall for more than two months. The shrinking of the river hurt the nearby residents and even the nation's economy. More than half of US grain exports are transported by barge on the river. The reduced channel depths and large exposed sandbars curtailed the flow of goods by barge by at least 45 percent. Access to clean drinking water was also put at risk as salt water from the Gulf of Mexico moved up the receding mouth of the Mississippi River.

The United States is not the only country that has been suffering from low river levels. Essentially all rivers in Europe were at low levels in 2022 with some dangerously low.[28] The Rhine River in Germany was especially low. Even the largest river in the world, the Amazon, remains exceptionally low because of the ongoing major drought in Brazil. Not only will these low levels impact river commerce and subsequently the economy in each country but also their agriculture from the lack of water for irrigation. The low river levels will also impact drinking water supplies in many areas. These are all very concerning climate change impacts that affect all people throughout the world with extremely few benefits. These are just a few of them. Climate change can surely not be seen as a positive development by anyone.

The people, including scientists and many writers, who have decided that climate change is irreversible and are happy just to slow down the trend are also not appreciating the severity of the problem. The situation is already very serious and needs to be stopped and reversed. This group of people might be considered to be with the people who think they can ignore climate change or expect the government to fix it with no pressure from the public. The government historically has rarely taken on any project that does not generate income or was in the interest of benefactors and lobbyists without significant public pressure. This unfortunate governmental stance has only gotten worse over the years. These two groups of people, however, recognize the problem and are in agreement that action needs to be taken to address it. This is a positive development.

The final group of people regards climate change in a doomsday scenario. With the daily reports of negative climate developments and predictions in the news and in social media, it is not unexpected that this would happen and it is easy to sympathize with them. However, as will be shown in this book, there have been several very severe instances of environmental pollution in the past and even present over which humans joined together and made focused efforts to resolve them. It was public pressure that forced the politicians and industries to resolve them regardless of the cost or political affiliation. It is the efforts and attitudes behind those victories that will be required to battle the climate crisis and prevent a planetary disaster. It is these victories that provide the confidence that the human race can prevail in this challenge.

THE TAKEAWAY

The climate crisis is far from a point of agreement for Americans and most other people worldwide. There are even people who regard it as potentially having positive outcomes. However, the data are all in agreement that the earth's atmosphere and ocean surface is warming at an alarming rate. This warming is in lockstep with increasing CO_2 levels in the atmosphere, which are now higher than they have been in the past 800,000 years and continue to rise at an alarming rate. There is no question that there are many negative and dangerous impacts from climate change including increased wildfires, increased and more destructive hurricanes and typhoons, rising sea level, melting glaciers, and reduced availability of surface fresh water among many others. There is no question that humans and, in particular, that the general public will need to take a stand and action to resolve this problem. This book will show how humans have faced many serious environmental threats in the past and have been victorious over them by taking the right approach. It is by using these attitudes and general approaches that the climate crisis can also be overcome in relatively short order. Without them, the people who worry that climate change will cause the downfall of human civilization will be correct.

Chapter 2

Rachel Carson, DDT, and Banned Pesticides

One of the most impressive examples of humans addressing and thoroughly remediating a serious environmental problem was spearheaded by Rachel Carson, the pioneer of the American environmental movement.[1] Her most concerted environmental crusade was to bring attention and action to the dangers of persistent organic pesticides which were widely used in agriculture to increase food production as well as in residential and even natural settings.[2] She termed them "biocides" because they killed so many forms of life in the natural environment upon contact. In addition, they are very slow to break down through natural chemical reactions and so persist long enough to permeate soil, air, and water as well as occurring in many plants, bacteria, organisms, and other forms of food. They also accumulate in many organisms damaging all predators higher up in the food chain. As a result, they wound up in basically every food and dwelling, causing worldwide environmental and human health damage for many years. She identified a group of persistent pesticides that she wanted to be banned but the most infamous was DDT. Several of the other related pesticides were even more toxic than DDT but DDT was used so extensively that it presented as a much more dangerous threat. Carson's efforts would lead to the establishment of the US Environmental Protection Agency (EPA) and the banning of the pesticides she identified including DDT, though she would not live to see those results.

DDT DEVELOPMENT AND BENEFITS

DDT is dichloro-diphenyl-trichloroethane and likely the most infamous of all pesticides. It was invented in 1874 by an Austrian chemist named Othmar Zeidler and studied by several others over the years.[3] Its effectiveness as an insecticide was discovered by Swiss chemist Paul Müller but not until

1939. This discovery was rewarded with the Nobel Prize in Physiology and Medicine for Müller in 1948. At first, DDT was just used to control insects that attacked fruits and vegetables. However, it was soon found to be a potent weapon against insect-borne diseases. As a result, the US military began experimenting with it to aid in operations during World War II. In one experiment on typhus-carrying lice, a developing typhus epidemic in Naples, Italy, was completely eradicated in just two months in 1943. It worked equally well indoors on malaria-carrying mosquitoes. Within months, the Army started spraying many building interiors worldwide. By 1944, they experimented with applying DDT outdoors to the whole town of Castel Volturno in Italy with great success.[4]

The word spread quickly about this miracle substance that protected service persons from these terrible diseases. As a result, most soldiers and sailors began carrying cans of DDT powder to kill bedbugs, lice, mosquitoes, and any other insects they encountered. Millions of DDT aerosol bombs were used in tents, barracks, and mess halls as well as in European refugee camps and jungle battlefields in the Pacific theater. It is estimated that DDT saved tens to hundreds of thousands of human lives during the war. After the war, survivors lined up in the streets of Europe to be doused all over with DDT in an attempt to evade dangerous diseases.

DDT has several negative health impacts to humans with exposure.[5] It produces excitability, tremors, and seizures with increasing dosage and can damage the liver. However, most effects dissipate quickly and the risks were deemed well worth it considering the benefits. DDE (dichlorodiphenyldichloroethylene) is an environmental metabolite of DDT that is naturally produced during the breakdown of DDT, and can reduce lactation and cause premature births which is a bit more serious. However, even though this health issue was known, it did not restrict the growing application of DDT.

When World War II ended in 1945, DDT was made available to the general public and its usage ballooned. Production of DDT in the United States increased from 4,366 tons in 1944 to 81,154 tons in 1963, a near eighteen-fold increase.[6] Many organizations like the US Public Health Service, the Tennessee Valley Authority, and the Rockefeller Foundation funded large-scale DDT projects to control malaria in many areas. DDT was used everywhere. It was sprayed in buildings and homes, from crop-dusting airplanes and out of the backs of trucks up and down the streets of many urban and suburban areas across the United States. Children would chase behind these trucks so they could be in the DDT mist. It was extremely effective. The number of malaria cases in the US dropped from about one million per year before DDT usage to 437 by 1952, a 99.99 percent decrease. It was applied just as pervasively across Europe and Canada, quickly eradicating malaria-bearing mosquitoes and typhus-bearing lice during the 1940s and early 1950s. The result was that human mortality rates were reduced twenty-fold in many areas.

RACHEL CARSON AND THE
TARNISHING OF THE IMAGE

Once DDT was proven to be so effective in 1939, research immediately began to develop several other persistent organic pesticides.[7] Chlordane is one of these new pesticides. It is volatile, evaporating quickly in air, which poses the risk of poisoning by inhalation to anyone applying it or exposed to it. It is very persistent in soil, food, and on surfaces. Ingesting just 2.5 parts per million (ppm) of chlordane in any food can lead to storage of 75 ppm in an organism's fat cells through the process of bioaccumulation. Heptachlor is another persistent pesticide and a constituent of chlordane. It is even more prone to bioaccumulation in fat cells with as little as 0.1 ppm being dangerous for consumption. The pesticide Dieldrin is five times as toxic as DDT through ingestion and forty times as toxic if absorbed through the skin. Exposure quickly causes convulsions in its victims. Aldrin is similar to dieldrin but even more toxic. It causes liver and kidney damage in humans and it is absolutely devastating to birds including domesticated fowl. Endrin is likely the most toxic of these pesticides. It is similar to Dieldrin but up to twelve times as poisonous to rats. It has killed cattle that have inadvertently wandered into sprayed orchards. These are just some of the group of newly developed persistent organic pollutants, many of which are pesticides. Through observational and some laboratory research, Carson identified twelve of these very dangerous pollutants including those already named and, in addition, toxaphene, Mirex, hexachlorobenzene (HCB), polychlorinated biphenyls (PCB), dioxin, and furans.[8] This list would later be adopted by the US Environmental Protection Agency as high priority for elimination. Nonetheless, production of the pesticides in the United States increased from 124,259,000 pounds in 1947 to 637,666,000 pounds in 1960, a fivefold increase.

The problem soon arose that, in addition to pests, these pesticides killed, birds, amphibians, many reptiles, some mammals, and all insects either directly or by killing off their food/prey. Many wild birds, frogs, and bats survive on the mosquitoes that were being exterminated and their populations plummeted as a result. Also, some insects like bees, butterflies, and other pollinators are beneficial as are ladybugs and praying mantises which prey on many damaging insects also suffered dramatic reductions in populations. Overencouraged and overzealous use of DDT and related pesticides resulted in human and wildlife poisoning incidents that caught the attention of the talented scientist and writer Rachel Carson. She studied several of these incidents in the field and through scientific reports and reported them through the media which caught the attention of the general public.[9]

In 1957, Carson reported an incident in which hundreds of songbirds in Massachusetts were poisoned and killed from aerial spraying of excessive amounts of DDT for mosquitoes. Later she reported on an incident where a DDT–fuel oil mix that was sprayed over Long Island, New York, poisoned a group of farmworkers. In several southern states pesticide use on fire ants also poisoned workers. In 1959, consumption of cranberries was banned for Thanksgiving because pesticides had been overapplied in the cranberry bogs. These incidents were the foundation for the writing and release of Carson's famous book *Silent Spring*.[10] This book was an immediate top seller and remained on best-seller lists for years. It clearly established Carson as the leader of the American environmental movement. It was declared as one of the twenty-five greatest science books of all time by *Discover* magazine. It is considered as impactful as *Uncle Tom's Cabin* by Harriet Beecher Stowe and *On the Origin of the Species* by Charles Darwin. *Time* magazine declared Carson as one of the most influential thinkers of the twentieth century. In 1992, a US Congressional panel declared *Silent Spring* the most influential book of the past fifty years.

As a result of Carson's notoriety, some of her writing was included in the Congressional Record.[11] In 1962, she received an award from Stuart Udall, the US Secretary of the Interior. President Kennedy became interested in Carson's findings and requested his Science Advisory Committee to investigate them. In 1963, Rachel Carson was introduced as the pioneer of the environmental movement when she testified to the US Congress. As a result of her advocacy, at least forty bills to regulate pesticide use were introduced in many states. On April 3, 1963, Carson appeared on national television to explain the dangers of pesticides. Unfortunately, Carson died of breast cancer in 1964 at the height of her influence. Many believed that her cancer resulted from exposure to chemicals she identified as dangerous.

Even at the height of her influence, Carson faced significant obstacles. Chemical companies hired scientists to disprove her findings and discredit her as a scientist. They attacked every aspect of her life, including personally, and accused her of mass murder for all of the people who would die from not applying DDT. However, all of the notoriety served to cement her position as the leader of the American environmental movement.

POST-CARSON PROBLEMS AND SOLUTIONS

Some of the scientific findings about DDT and related pesticides that were made in the 1950s blossomed as more serious problems in the 1960s. These pesticides strongly adhere to the soil particles and are highly persistent in the environment.[12] In warm temperatures, they can evaporate or be carried in the soil particles by the wind. As they are condensed or the impacted dust settles, they may have traveled thousands of miles from where they were applied and this cycle of erosion and deposition can happen many times before the pesticide degrades. In this way, these persistent pesticides impact a much larger area than just the application site. For example, DDT has been found as far away as the Arctic. When DDT is finally degraded by microorganisms, it can produce DDE, which is also dangerous, and it can take fifteen years for this to happen.

Bioaccumulation

DDT penetrates the skin if dissolved in oil. If ingested, it is also absorbed in the digestive tract and it is absorbed by the lungs through inhalation. DDT is soluble in fat and is stored largely in fatty organs in the body such as the adrenal glands, testes, and thyroid, as well as the liver, kidneys, and various areas of body fat. Storing DDT in organs and fatty tissue biologically magnifies it.[13] Intake of 0.1 ppm in the diet concentrates to 10–15 parts per million in fatty tissues. This process is called biological magnification which can lead to health issues and even poisoning. It can also accumulate in humans through the food they eat. If hens eat grain with DDT, the eggs they lay will also contain DDT. Further, bigger and fatter animals bioaccumulate pesticides to higher concentrations than short-lived, smaller animals. Under the same conditions, a large old trout bioaccumulates much more DDT than a young bluegill. If cows eat hay containing a few parts per million of DDT, it will concentrate in the milk they produce at about three parts per million. If butter is made from this milk, it can have as much as 65 parts per million.

However, these pesticides were in most foods and at higher levels than 0.1 parts per million. One study found that from 1971 to 1976, 96 percent of commercially available milk in Illinois contained dieldrin, 93 percent had heptachlor epoxide, 73 percent had lindane (another persistent pesticide), 69 percent had chlordane, and 48 percent had DDT.[14] Other studies found that 99 percent of packaged milk contained DDT and other pesticides. The problem was that when lactating mothers consumed dairy products from this impacted milk, the pesticides entered their bloodstream and were transferred to their milk without breaking down. Infants then consumed this pesticide-ridden breast milk.

This problem had been known for a long time. In 1951, a study found that 30 of 32 tested lactating African American women from Washington, DC, had DDT in their breast milk with an average of 0.13 parts per million.[15] However, no one investigated this again until 1965 and then multiple studies were conducted over the following years. A large study of more than 1,400 American women was released in 1981 and it found that 83 percent of the test subjects' milk contained dieldrin, 74 percent contained chlordane (oxy), and 61 percent contained heptachlor (epoxide).[16] A similar study in Canada in 1981 found that 99 percent of the human mothers' milk tested contained DDT, 98 percent contained aldrin, and 80 percent contained the DDT metabolite DDE.[17]

The amounts of DDT in human breast milk were also problematic. Some studies found average concentrations of more than 1.0 parts per million which is far above recommended maximum. Table 2.1 shows typical and allowable concentrations of dieldrin, heptachlor, and DDT in breast milk as well as the maximum concentrations in cow's milk as determined by the US Food and Drug Administration.[18] All of the research studies found that human breast milk had DDT levels far in excess of allowable levels. It was for this reason that new mothers were advised not to nurse their infants but instead to feed them formula in the 1960s through 1980s.

Prenatal exposure to persistent pesticides and DDT in particular has been shown to result in newborns with shorter birth length, lower weight, and smaller head circumference. Further, the growing impacted children showed delayed social and motor development skills and a higher occurrence of attention deficit/hyperactivity disorder.[19] With continued studies, it was found that long-term exposure increased the incidence of high blood pressure, Alzheimer's disease, reproductive disorders, possible breast cancer, and other conditions. It even caused genetic damage which impacted two or three generations after exposure. However, these threats were confounded by other published studies that claimed there was no danger from DDT or the other pesticides.

By the mid to late 1960s, DDT and the other pesticides became a flash point of controversy.[20] Not only was there the issue with pesticides in milk and foods but it was causing significant damage to the environment. Public pressure in reaction to *Silent Spring* and the awakening to the damaged environment was overwhelming as was the demand to ban DDT. The United States government was forced to establish the Environmental Protection Agency (EPA) in 1970 by public pressure. One of the EPA's early acts was to ban DDT and the other pesticides identified by Carson in the United States

Table 2.1. FDA Action Levels, Allowable Intake and Nursed Intake of Pesticides

Pesticide	FDA Limit—Cow Milk (ppb)	Allowable Intake (mg/kg)	Typical Mothers Milk (mg/kg)
Dieldrin	7.5	0.1	0.8
Heptachlor	7.5	0.5	4
DDT	50	5	28

in 1972. The first EPA administrator, William Ruckelshaus, attempted to oppose the bans but public pressure was overwhelming and he succumbed to supporting it.

As a result of the ban of DDT and other pesticides in the United States and Canada, DDT in human milk fat dramatically declined.[21] Concentrations can also be expressed in μg (microgram) of pesticide per kilogram of containing material. Concentrations of DDT in breast milk declined from 5,000–10,000 μg of DDT per kilogram of milk fat before the ban to less than 500 by 1990 in the United States and Canada. Other industrialized countries and areas that also banned DDT saw similar declines over about the same period. Some areas, however, did not ban DDT and the women and infants continued to be exposed. Further, the bans do not ensure that people in the countries with bans are completely free from exposure. Food imported from countries where DDT and the related pesticides are not banned commonly still contains them. Consumption of that food has been exposing people to these older pesticides though the amounts have been in steady decline with time.

Biomagnification

There is still debate over how damaging low doses of DDT are to human health and whether banning it helped or hurt more people. However, there is no question that banning it improved the quality of the natural environment. There was no debate that DDT and the related pesticides are extremely

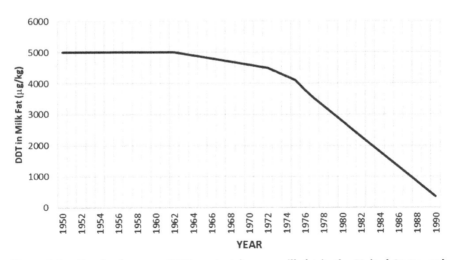

Figure 2.1. Graph of average DDT content in cow milk fat in the United States and Canada from 1950–1990.
Source: Adapted from Smith (1999).

destructive to insects, birds, reptiles, amphibians, and some mammals through both direct poisoning and reduction of food supplies for them. However, like bioaccumulation, there is another long-term danger related to the persistence of these pesticides. This danger is biomagnification.

In *Silent Spring*, Rachel Carson described the mechanism of how the persistence of these pesticides allowed them to be ingested by organisms at the bottom of the food chain or food web and passed up to higher-level organisms through predation without chemically breaking down.[22] Because these pesticides accumulate in certain parts of the organism, the predator will have higher amounts of the persistent pesticides in their organs and fat deposits that accumulate pesticides. This step-by-step process focused accumulation in an organism with time and the predation of that group of organisms by the next higher organism in the food chain means that the pesticide increases in many higher food chain organisms. For example, a study of the biomagnification of DDT found that if DDT was at a concentration of 10 parts per million in soil, earthworms that consumed the soil had more than 140 parts per million of DDT and robins that consumed the earthworms had over 400 parts per million. Certainly, by the time the DDT and other pesticides reached the top of the food chain or food web, such as humans, these concentrations could be extreme and cause severe damage to the organism.

One group of organisms that was particularly impacted by pesticide biomagnification during the 1950s and 1960s was raptors or birds of prey. They are the top of their food chain and therefore prone to the maximum concentrations of persistent chemicals and resulting ill effects. In particular, the bald eagle suffered greatly from DDT and persistent pesticide biomagnification.[23] At the time that the bald eagle was designated the symbol of the United States, in 1782, there were at least 100,000 nesting eagles in the United States. Earlier that century, the bald eagle population in the United States was likely to have ranged from 300,000 to 500,000 birds. The problem is that they were regarded as nuisances at that time and deemed as threats to livestock and salmon stocks. Bald eagles were commonly killed by poisoning or shot to reduce their numbers. Alaska even offered a 50-cent bounty per bald eagle carcass, and later increased it to one dollar in 1917. These actions resulted in more than 120,000 confirmed deaths of bald eagles. By 1940, so many bald eagles had been exterminated that the US Congress passed the Bald Eagle Protection Act which outlawed the disturbing or killing of bald eagles or even possession of any eagle parts, including feathers, eggs, and nests.[24]

When DDT and the other pesticides were introduced in widespread agricultural use, they were adopted largely to control many crop-damaging insects such as tobacco budworms, Colorado beetles, flies, potato beetles as well as Dutch elm disease. Farmers applied DDT in agricultural areas by hand as well as using airplanes and helicopters. DDT was sprayed from trucks on most city streets around the United States and other industrialized countries and even applied directly to people. These activities spread DDT throughout the environment on a massive scale.[25]

This applied DDT was washed by precipitation to surface-water runoff and eventually to rivers, ponds, and lakes where it entered the water cycle. DDT deteriorates reasonably quickly in sunlight in the air but in soil, DDT has a half-life of fifteen years. A half-life is the time needed for one half of the mass of the pesticide to chemically break down into other less dangerous chemicals meaning that DDT can last for decades in soil at full potency. The DDT in the surface water was first ingested by plants and microorganisms which in turn were ingested by small predators which in turn were ingested by increasingly larger predatory organisms.[26] Most DDT entered the food chain through aquatic plants which were eaten by fish which, in turn, were eaten by increasingly larger and more predatory fish which were finally consumed by bald eagles and other apex predator birds. The DDT bioaccumulated in the eagle as it ate fish rich in DDT through biomagnification. This results in thousands of times the DDT in eagles than was in the primary producers.

Bald eagles are highly susceptible to biomagnification of pesticides because they are large predators, live at least twenty-five years, cannot reproduce until at least five years of age, and their diet is primarily upper-food-chain fish. They consume a large amount of DDT from these fish and it accumulates over their long lifetime. Osprey, peregrine falcons, brown pelicans, and many other large seabirds and waterbirds are at similar risk for the same reasons of bioaccumulation and biomagnification. They are also apex predators that consume a large amount of upper-food-chain fish so ingest and bioaccumulate DDT and other persistent pesticides. Research studies found a large amount of DDT metabolites in adult eagles but no correlation with mortality from exposure to DDT. The DDT was not killing adult eagles. This was very confusing because there was a sharp decline in the eagle population through the 1950s.[27] By 1963, there were only 412 breeding pairs of bald eagles identified in the United States in total. After extensive research, it was found that DDE, the environmental metabolite of DDT, was the cause of the decrease in the eagle population. Laboratory experiments confirmed field observations that DDE causes pronounced thinning of the shells of eagle eggs. DDE exposure makes absorbing calcium difficult for the birds. The low calcium levels result in thinner-than-normal eggshells. As a result, female eagles were crushing their eggs when they sat on them for incubation. The vast majority of eagle eggs were crushed before they could hatch. Field studies confirmed that exposure to DDE was sufficient to cause the massive decline in population.

BAN OF DDT AND RECOVERY OF BIRD POPULATIONS

Bald eagles were added to the endangered species list in 1967 but many people believed that they would quickly go extinct.[28] The federal protection efforts halted the rapid decline of bald eagles but the populations also did not recover. However, this situation changed in 1972 with the ban on DDT in the United States. Because of the ban, newly introduced successful breeding programs, reintroduction of captivity-bred eagles into the wild and the poaching protection from the Endangered Species Act, the bald eagle population rebounded as quickly as it had fallen. The recovery was so dramatic that in August 1995, the bald eagle was removed from endangered status to threatened status. By 1996, more than 5,000 breeding pairs of bald eagles were identified. By 2007, the recovery of bald eagle populations was so great that it was completely removed from federal protection. A 2009 survey identified 72,000 bald eagles and about 30,000 breeding pairs in the United States. By the fortieth anniversary of the banning of DDT in 2022, there were 316,700 bald eagles including 71,400 nesting pairs in the lower 48 states. This is essentially a complete recovery of a species that was all but extinct purely by addressing the problem.

Many other predatory and scavenger birds benefited from the pesticide bans as well. Saving the California condor became a battle cry of West Coast environmentalists during the 1960s and it experienced the same magnitude of recovery as bald eagles after the ban of DDT. Ospreys experienced similar recoveries but perhaps the most impressive raptor recovery was the peregrine falcon.

Before the 1940s, there were about 7,000 peregrine falcons including 3,875 nesting pairs in North America, primarily in Alaska and Canada.[29] There were more than 8,000 falcons in Europe. The decline in peregrine falcon population in the United States began in the 1940s but was most pronounced from 1950 to the early 1970s. Falcons suffered the same inability to incubate thinned eggs as bald eagles but were also fatally poisoned as a result of the biomagnification and bioaccumulation. By the time action was initiated, the American falcon population was down to 12 percent of that prior to pesticide and DDT introduction. In 1970, the peregrine falcon was placed on the Endangered Species List. By that point, the birds had been completely eradicated in the Midwestern and Eastern United States through exposure to DDT and related pesticides, and the Western populations were almost 90 percent below historical levels. In 1975, there were 324 breeding pairs of peregrine falcons in the continental United States with one count of only 39 pairs and all were in the west.

After DDT and related pesticides were banned in 1972, several extraordinary efforts by federal and state Fish and Wildlife Services, the Cornell University Peregrine Fund, the Midwestern Peregrine Falcon Restoration Project, the Canadian Wildlife Service, and many other dedicated groups and individuals restored the falcon population. More than 6,000 falcons were raised in captivity in several efforts and released to the wild in North America. By 1996, there were 993 nesting pairs of falcons in the lower 48 states and by 2006 it was up to 3,005. As a result of this rapid recovery, on August 25, 1999, the US Fish and Wildlife Service delisted the American peregrine falcon from endangered and threatened species designation. They considered this among the most dramatic successes of the Endangered Species Act.

The problem was not just with the raptors. The brown pelican was driven to near extinction largely as the result of exposure to DDT and related pesticides and it too made a phenomenal recovery.[30] The birds already had trouble before they were impacted by pesticides as their feathers became fashionable in garments around 1900 and many thousands were killed. Later in 1914, fishermen shot and killed many pelicans because they were believed to be reducing the fish populations. It took until 1918 with the passage of the Migratory Bird Treaty Act for the killing of these migratory birds to become illegal. Even with these attacks, the brown pelican populations were still healthy with Texas alone having at least 50,000 and as many as 85,000 birds in the 1930s.

Just as with the other examples, the introduction of DDT and related pesticides in the 1940s devastated the brown pelican populations through the early 1970s. Louisiana made the brown pelican its state bird in 1966 but they were already extinct in the state by that time. By 1961, Texas had less than 50 birds from their huge numbers just 30 years earlier. Florida had just 7,690 brown pelican nests in 1970 when the bird was finally listed as an endangered species. By then, Louisiana had already begun recovery efforts reintroducing 1,276 young pelicans to three sites in the state in 1968. Brown pelicans in California, however, continued to decline, reaching a low of 466 pairs in 1978. They had the same problems as raptors of weak eggshells and outright poisoning.

As with the other examples, once DDT and related pesticides were banned and the remaining birds were protected, populations quickly recovered. Brown pelicans along the Atlantic Coast and in the South recovered so quickly, they were removed from the Endangered Species list in 1985, just fifteen years after they were added. By 1999, the population of pelicans in Louisiana had increased from zero to about 50,000. By 2009, the recovery was so great that the US Department of the Interior removed the brown pelican from the protected species list. At that point, the global population of brown pelicans was about 620,000 with about 172,000 along the California and Gulf of Mexico coast. At that time, there were 11,695 nesting pairs in California from the 466 just 31 years earlier. It is estimated that Texas and Louisiana had more brown pelicans than ever in history.

THE TAKEAWAY

Whether DDT and the related persistent pesticides are viewed as boons to the human race or scourges of the natural environment, the impressive aspect is that a grassroots movement started largely by one person, Rachel Carson, was able to cause the banning of these substances and reverse most of the damage they had caused in several important areas. The president of the United States at the time of the ban was Richard Nixon, who was in no way a friend to the environment, and he opposed the ban. However, public pressure was so strong that he and his appointees as well as other elected officials had no choice but to comply. This is a prime example of a human triumph over pollution completely instituted by a governmental change in policy. It required cooperation and a concerted effort to pressure public officials to act. Individual activists could only abstain from using pesticides around their homes which would have had only a minor impact on the overall effort. This is not always the case. However, it shows that elimination of chemicals from the environment that are essentially ubiquitous and the recovery of impacted species and areas by them are possible and effective with effort by concerned individuals.

Chapter 3

Get the Lead Out

A surprising environmental crisis of the twentieth century that was addressed and largely eliminated in a spectacular manner was, and is, environmental lead. Recently, the dangers of lead are only heard about when legacy usage becomes a current issue. Old public water pipes in Flint, Michigan, and home plumbing in Newark, New Jersey, have become alarming and surprising situations recently. However, lead is ranked as the second most dangerous environmental pollutant on the 2019 US Comprehensive Environmental Response, Compensation, and Liability Act (CERCLA) priority List of Hazardous Substances with only arsenic ranked as more dangerous.[1] Its negative health effects from exposure have been known for millennia and yet, during the twentieth century, it became among the greatest threats to public health and the environment. It was largely through the efforts of one person that this developing disaster was averted and resolved effectively, forever.

WHAT IS LEAD?

Lead is a naturally occurring, inorganic element that is classified as a heavy metal.[2] It occurs as a natural part of many rocks, minerals, and soils throughout the world. It was likely the earliest metal to be mined and utilized by humans for tools, weapons, and everyday objects. Its low melting temperature and malleability relative to other metals and materials make it easy to smelt, refine, and work into usable objects. The most common natural occurrence of lead is in the mineral galena which is lead sulfide. Not only did and does the mining of galena create extensive surface pollution, the air pollution fallout of sulfuric acid from smelting it has damaged an even broader area throughout history.[3] However, this localized impact was minor compared with the human and environmental poisoning from the lead itself.

The oldest identified lead mine is in Turkey and probably from about 6500 BCE.[4] The oldest lead artifact so far identified is from necklace beads found in Anatolia, Turkey. Total tonnage of mined and manufactured lead appears to have been relatively low until the invention of the process of cupellation where silver is separated from lead. As a result, production was greatly increased. Production continued to increase steadily thereafter partly because of the production of coins which expanded in about 750 BCE but lead usage peaked during the Roman Empire through about 100 CE when many of the more productive mining operations were expended. The Romans had a huge smelter in Spain that was operated by tens of thousands of slaves and another major lead smelting site in Greece.

The Romans had a love affair with lead, producing an average of 66,140 tons (60,000 mt) of lead per year for about 400 years. Lead was used in cosmetics including face powders, rouges, and mascaras.[5] It was used as pigment in many paints which poisoned the painters causing them to exhibit abnormal behavior and sometimes causing them to develop serious mental health issues.[6] Lead acetate was used as a sweet and sour condiment for seasoning food and a wine preservative and sweetener to disguise the flavor of poor vintages. A commonly used grape-based sweetening syrup contained excessive amounts of lead. Of the 450 recipes in a popular ancient Roman cookbook, 100 of them included these syrups as ingredients. Lead was used as an ingredient in pewter cups, plates, pots and pans, pitchers, and other household tools as well as vinting vessels to produce most wines.[7] It is estimated that Roman aristocrats consumed between one and 5.3 quarts (1–5 l) of tainted wine per day. Lead was also used as a component in the production of metallurgical alloys in bronze and brass coins as well as counterfeit silver and gold coins.[8]

The main use of lead, however, was in the production of water and sewage pipes for the extensive network of plumbing throughout the Roman Empire. All drinking water in Rome and the other major cities of the Roman Empire was supplied using lead pipes. The word "plumbing" is from the Latin word for lead and "plumber" means worker of lead. This meant that virtually every urban resident of the Roman Empire was exposed to excessive levels of lead.

LEAD EXPOSURE AND POISONING

The extensive use of lead led to a fourfold increase in lead in air pollution during the Roman Empire. Its negative health effects were well-known during this entire period. The debilitating effects of chronic lead poisoning were described as early as the seventh century BCE. Roman writers described the toxicity of lead in public health. The serious health effects included severe mental illness and even death but the many uses of lead were so popular that the government and most writers minimized the hazards it posed. As a result, it was one of the major factors in illness and death throughout the Roman Empire. Children were especially impacted by lead because they had high exposure levels and were more affected by the health consequences.[9] It is estimated that drinking water from Roman plumbing likely contained up to 100 times more lead than spring water as determined through research on the chemistry of sediments of that age.[10] Human skeletons from across the Roman Empire were found to contain excessive levels of lead. It was even believed at one point that lead poisoning from plumbing and food was a contributing factor to the fall of the Roman Empire. It was thought that lead poisoning may have weakened the public health of the Romans so they couldn't defend themselves against attack. This theory, however, later fell out of favor.

Lead use declined somewhat in the early Middle Ages but increased again before the Renaissance.[11] The first increase corresponded to increased silver production in Germany but it was not exclusive to silver. With the increased availability of lead, it again became a public health crisis. In parts of Germany, lead poisoning was so widespread that wine producers and distributors found mixing lead sugar syrup into their wine could be sentenced to death. The laws were first enacted in 1498 in some towns and later in 1577 in others. However, the lead poisoning plague occurred in many other areas.[12] After a particularly widespread outbreak of stomach illness as a result of lead in wine, in 1696, Duke Ludwig further extended the prohibition of the use of leaded additives in any wine product under the penalty of death. The epidemic was not confined to wine. Lead was also leached out of the glazed pottery used to store beer which resulted in gastrointestinal disease and chronic colic outbreaks in Germany during this period.

The lead poisoning plague was not confined to Germany. In 1763, a physician in King George III's court determined that lead fittings in cider presses caused a widespread outbreak of colic and stomach problems in England.[13] Throughout eighteenth-century England, there were widespread gout epidemics that were later traced to popular port wines from Portugal. In 1825 alone, more than 5.5 million gallons (21 million l) of port was consumed in England. These ports were heavily leaded. This poisoning plague also spread to America at this time.

By the twentieth century, the United States had become both the world's leading producer and consumer of lead.[14] Despite the problems that lead plumbing had given the Romans, it was extensively used for the water mains as well as home plumbing. Even when copper pipes replaced the lead pipes, they were soldered together using lead. This ensured that lead was present in drinking water throughout the United States and most of the industrialized world.[15] Another major use of lead was in paint. This use began in the early nineteenth century when lead carbonate was termed "white lead" and used as a pigment and primary component in white and pale colored paints. Although it was banned in Australia in 1914 and in an international convention in 1925, before 1955, it made up as much as 50 percent of the volume of white paint used in the United States and many other countries. In addition, lead orthoplumbate and lead oxide was termed "red lead" and used for red, pink, and orange pigment through the 1960s. It is estimated that 83 to 86 percent of houses built before 1978 contain lead paint.[16] However, the primary use peaked in the 1920s and progressively decreased until it was finally banned.

By the 1920s, automobiles were becoming more sophisticated and began experiencing problems resulting from the fuel properties. They experienced excessive preignition or knocking that reduced their performance and limited their advancement. In an attempt to overcome the situation, researchers added tetraethyl lead (TEL), an old German patent, to gasoline in 1921. This greatly improved the performance of engines in laboratory testing and it was quickly adopted for commercial use as a new fuel called ethyl gas.[17] However, the danger of this new mixture was quickly realized. By 1924, there had been at least thirteen to fifteen deaths and more than 300 cases of men becoming psychotic from working with tetraethyl lead.[18] As a result, in May 1925, the surgeon general of the United States suspended the production and sale of leaded gasoline and convened a committee to study the problem. The committee released a report in June 1926, finding no solid reasons for prohibiting the production and sale of ethyl gasoline. In response, the surgeon general issued a voluntary standard to the oil industry of 3 cubic centimeters of tetraethyl lead per gallon of gasoline. This standard was later raised to 4 cubic centimeters of tetraethyl lead per gallon of gasoline in 1958. Testing during the 1950s and 1960s found that the actual amount of tetraethyl lead per gallon of gasoline was about 2.4 cubic centimeters per gallon. Ethyl gas was, by far, the greatest source of environmental lead emissions, accounting for as much as 90 percent of exposure.

By the twentieth century, the United States had become both the leading producer and consumer of refined lead in the world.[19] By 1980, consumption was about 1.3 million tons (1.2 million mt) of lead per year or about 40 percent of the total world production. Spread across the population, this consumption translated to 5,221 grams of lead per American per year. About 85 percent of this lead was produced domestically and about 40 to 50 percent was the result of recovery and recycling efforts. By then, about 88 percent of this lead was from just seven mines in the New Lead Belt, Missouri.[20] In addition to paint, plumbing, and gasoline, lead was used in a number of common applications. These included stained glass, jewelry, pottery, shot and bullets, fishing sinkers, balance weights on car wheels, electrical and electronic devices, automotive batteries, pewter, some alloys, and radiation shielding, among others. Even canned foods were a source of lead exposure because acidic foods, like tomatoes, leached lead from solder joints in the cans.

Lead poisoning was rampant especially at the beginning of the twentieth century but it was primarily related to occupational exposure.[21] Between 1875 and 1900, there were about 30,000 cases of lead poisoning in the lead mines of Utah alone. In Great Britain at this time, there were about 1,000 cases of poisoning per year with about fifty being fatal. Lead poisoning effects on children were documented in Brisbane, Australia, in 1892 which led to restrictions on lead paint in that country. Expanded studies on deleterious effects of lead on neurological development in children were released in 1943. They concluded that exposure to environmental lead resulted in behavioral problems and lowered intelligence. The reason is that lead replaces calcium in the body which interferes with how neurons communicate in the brain.

Later research provided a more comprehensive picture of the impact of lead exposure on humans.[22] One-time or acute lead poisoning at even moderate exposure produces sluggishness, nausea, vomiting, painful gastrointestinal issues, diarrhea, loss of appetite, colic, weakness, dehydration, and discoloration of the lips and skin. At high exposure, it produces convulsions, limb paralysis, coma, and finally death. Chronic or repeated exposure results in several physiological effects such as the previously identified neurological effects, renal damage, blood damage, endocrine damage, cardiovascular effects, hypertension, reproductive damage, developmental problems, and cancer. Lead exposure in children produces such profound health effects that the US Centers for Disease Control (CDC) developed a system of poisoning levels with recommended treatments. The primary area of damage from lead poisoning is the nervous system and children are more profoundly impacted than adults. Symptoms in children include decreased IQ scores, loss of memory, ADHD, depression, delayed reaction time, irritability, muscle tremors and weakness, decreased hand-eye coordination, hearing loss, and even early onset of Alzheimer's disease.[23] Lead also damages the kidneys including impairment of proximal tubular function, chronic nephropathy, gout, and kidney failure. Lead can decrease hemoglobin in the blood, and cause anemia. Lead negatively impacts cell growth and maturation and bone and tooth development. Lead exposure can also cause a significant increase in blood pressure. Lead exposure can cause decreased sperm counts and quality, as well as increased stillbirths and miscarriages, premature births, low birth weights, and birth defects. The US National Toxicology Program categorizes lead as "reasonably anticipated" to be a carcinogen.[24]

With all of the known detrimental health effects of lead for many centuries, if not millennia, it would be reasonable to assume that people would know better than to widely use it or release it to the environment. But in the nineteenth and twentieth centuries, when people were supposed to be better educated, the pandemic of lead became more widespread and dangerous than it ever had been.[25] The reason is that it was inexpensive and served a multitude of purposes with less effort than alternatives. As a result, there was widespread pressure from many industries to minimize regulation of lead despite the environmental and public health costs. Like DDT, it would take a strong leader to turn the tide and that leader was Clair (Pat) Patterson.

PATTERSON'S BATTLE AGAINST LEAD

Pat Patterson was born on June 2, 1922, in Mitchellsville, Iowa.[26] He graduated from high school in 1940 and graduated from Grinnell College, Iowa, in 1943 with a bachelor's degree in physical chemistry. He earned a master's degree in chemistry from the University of Iowa in 1944. His advisor invited Patterson to the University of Chicago to work on the Manhattan Project but after several months of work, he decided to enlist in the military. However, he was rejected because of his high security clearance and was transferred to the Oak Ridge, Tennessee, laboratory facilities to continue working on the Manhattan Project. There, he researched mass spectrometry of uranium isotopes.

When World War II ended, Patterson returned to the University of Chicago to earn a PhD. In 1948, he began his dissertation on lead isotopes and developed mass spectrometry methods to determine the age of rocks. He devised the laboratory procedures for this analysis and found that background levels of lead were extremely high. He was forced to develop procedures to keep the lab ultraclean or his samples would become severely contaminated by ambient lead. He graduated in 1951 but remained at the University of Chicago as a postdoctoral fellow. His advisor, however, moved to the California Institute of Technology the next year and Patterson went with him. There, he studied lead isotopes in meteorites and, in 1953, found that all meteorites are the same age. This research allowed him to make one of the greatest scientific discoveries of the twentieth century: that the earth and the entire solar system are 4.55 billion years old. Patterson was hailed as a star in the scientific community.[27] This momentous discovery was more than enough for a whole career but not for Patterson. Now a faculty member at Caltech he set about characterizing the lead contents and isotopes of common geological materials. This research led to his activism to reduce anthropogenic lead.[28]

Patterson's interest in lead problems began with his difficulty in building a "clean lab" when he was trying to study isotopic lead. He now used his clean lab to analyze minute quantities of lead in normal geological and anthropogenic materials documenting that there was a dangerous amount of lead in the modern environment. By 1962, Patterson documented that anthropogenic lead was being deposited in the natural surface environment at 80 times that of the rate in the oceans.[29] In 1963, he showed that water in lakes and rivers had three to ten times as much lead as deep ocean water even though it should have been much less. Then, in 1965, he published a study that showed Americans were being exposed to 100 times the natural background levels of lead and just short of outright lead poisoning on a national level. This report was in sharp contrast to the industry claims, that their activity increased public exposure to lead by less than two times normal background.

Patterson found these excessive lead levels in the blood of random volunteers. The major sources of this lead were found to be from gasoline, paint, solder in pipes and electrical devices, and pesticides. These findings were highly regarded by the scientific community because of Patterson's reputation as an eminent scientist. However, they were not so welcomed by industry or the government at all levels because addressing them would be costly and bad for business.

In retaliation, Patterson was attacked and ridiculed professionally and even personally by several respected industry professionals who cast him as an environmental fanatic and a rabble-rouser. There are even reports of Patterson having his life threatened. The negative press and threats, however, did not discourage him and instead, he redoubled his efforts. He wrote a letter on October 27, 1965, to then California governor Pat Brown and warned him about the danger to the public from airborne lead, especially in and around Los Angeles.[30] This effort, however, was politely rejected. He sent a letter warning about lead on October 7, 1965, to Senator Edwin Muskie, the then chairman of the federal Subcommittee on Air and Water Pollution. As a result, Patterson was invited to Washington, DC, to testify at a congressional hearing on June 15, 1966, where he presented the results of his research and issued a warning of the serious threat to public health. When he returned to California, on March 24, 1966, Patterson sent a second letter of warning to Governor Brown. This letter was better received and action was taken as a result. On July 6, 1966, the governor's office directed the California Department of Public Health to establish air quality standards for lead by February 1, 1967.[31]

These advocacy efforts could have distracted Patterson from research on anthropogenic lead, but instead, he increased it to better support his findings.[32] He published a study in 1970 on lead in ice cores that showed lead in current snow in Greenland had 100 times as much lead as snow from preindustrial times. He also showed that current snow in the Antarctic had tenfold more lead than in preindustrial times, largely because there is less land, fewer people, and less industry near the Antarctic. In spite of all of Patterson's findings, the National Research Council published a report on airborne lead that ignored Patterson's work meaning that even the scientific community did not fully support him. But even this slight did not dissuade Patterson from his mission.

A major advancement in this effort occurred in 1970.[33] Through massive public pressure, the federal government underwent a significant policy shift in its role on controlling water and air pollution. They developed and enacted the Clean Air Act of 1970 which authorized comprehensive federal and state laws regulating emissions from industrial and transportation sources. This legislation set the National Ambient Air Quality Standards (NAAQS) for six common air pollutants known as "criteria air pollutants" that (1) are widespread, (2) are harmful to public health and the environment, and (3) cause property damage. Likely, primarily as a result of Patterson's work, lead was included as one of the six criteria air pollutants. In addition, the Environmental Protection Agency (EPA) was signed into law on December 2, 1970, to implement the requirements of this and other environmental acts.

This new oversight federal agency might have signaled the culmination of Patterson's advocacy work on anthropogenic lead and his return to basic research but this still was not the case. He published a paper in 1973 showing that lead from gasoline emissions permeated the most remote wilderness. Possibly in response, later that year, in December 1973, the EPA announced a phased reduction of lead in gasoline by 60 to 65 percent. The restrictions were scheduled to begin on January 1, 1975, and to become increasingly more restrictive over the next five years. The average lead content of gasoline of each refinery would be reduced from approximately 2.0 grams per gallon to a maximum of 0.5 grams per gallon after January 1, 1979. There were some delays in initiation but it was carried out. Later, a ban on all leaded gasoline would begin in 1999.

There was also the Lead Contamination Control Act of 1988 which greatly restricted the release of lead to the environment. The EPA regulated lead in water through the 1972 (and later) Safe Drinking Water Act by establishing a 15 parts per billion (ppb) limit on lead in water and a goal of zero.[34] They further limited lead in air in public spaces to 1.5 micrograms per cubic meter. If 1 pound (0.45 kg) of lead arsenate or 10 pounds (4.5 kg) or more of any other forms of lead are released to the environment, the National Response Center had to be notified, which is still the case today. The Occupational Health and Safety Administration (OSHA) limits lead in workplace air to 50 micrograms per cubic meter for an eight-hour workday in a forty-hour workweek. If the blood lead level of a worker reaches 50 micrograms per deciliter or higher, then the worker must be removed from the workroom.[35]

Patterson continued his work despite these triumphs and showed, in 1975, that lead in high enough concentrations can overwhelm the natural defenses of plants that normally exclude lead from their tissues.[36] This means that lead can be incorporated into vegetables and other plants if it is at high enough levels in the soil. A very telling study was released by Patterson in 1979 where he analyzed the bones from 1,600-year-old Peruvian skeletons and compared them with modern human bones.[37] It showed that there was an increase in lead of 700 to 1,200 times in modern humans relative to humans from preindustrial times. This shows just how badly impacted modern humans had become.

Patterson was still not satisfied and took aim at lead in food from processing and storage methods. In 1979, he criticized the EPA for utilizing analytical methods that were not sensitive enough to properly determine the amount of lead in tuna. On October 10, 1981, Patterson publicly repeated this opinion at a symposium in Washington, DC, at which officials from the EPA and Bureau of Foods were present. Just a few months later, the EPA changed its standards to meet Patterson's recommendations. In 1980, Patterson was involved in research that showed there were excessive levels of lead in canned goods as a result of leaching from the solder in the lid seals. By 1991, lead had been removed from food cans, paints, waterlines, and gasoline, among other materials. As a result of all of this reduction, lead in new fallen snow in Greenland was of 7.5 times less than it had been in 1971. Patterson almost singlehandedly had enhanced public health.[38]

For one person to have such an impact on an environmental problem is rare indeed. There have been other environmental leaders such as Rachel Carson and Roger Revelle who identified a problem and brought it far enough to public attention that a movement began and action was taken. Patterson is unique as he identified a major environmental problem and carried it through to resolution. It is because of Patterson that gasoline, paint, pipe solder, and many other common items are now lead-free. The removal of lead additives from gasoline, which culminated in 1990, is regarded as among the major public health triumphs of the twentieth century. Between 1973 and 1994, the average lead concentration in American children's blood decreased from 13.7 mcg/dL to 3.2 mcg/dL.[39] Overall, it went from almost 16 mcg/dL in 1973 to about 1 mcg/dL in 2006. This decrease was in direct proportion to the amount of tetraethyl lead produced. This impact was not only in the United States. In the United Kingdom, in the early 1980s, about 7,000 tons of lead per year was emitted from automotive exhaust but by 2000, it had been reduced to just 30 tons per year. These are phenomenal changes and demonstrate that humans can overcome very significant environmental issues with the right approach.

THE BUNKER HILL LEAD HOT SPOT

Although the average person's exposure to lead has decreased phenomenally, there have been lead exposure hotspots from legacy usage and pollution that have caused significant local impact on public health. These hotspots largely resulted from mining, processing, and usage. A good example of a highly polluted mining operation is the Bunker Hill Mine site in Idaho.[40] Mining began

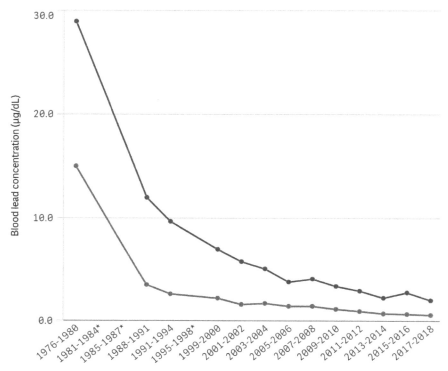

Figure 3.1. **Graph of lead concentrations in blood from children from the United States one to five years of age from 1978 to 2016. The top (black) curve is the maximum level (95% confidence) and the bottom (gray) curve is the median blood lead level. The limit for lead in blood was set as 5 µg/dL by the World Health Organization (WHO) and 3.5 µg/dL by the US Centers for Disease Control (CDC). The average level for children dropped below the CDC threshold in 1990 and the maximum level dropped below the threshold in 2010 never to increase.**
Source: EPA.

in 1883 for lead and silver but it was expanded in 1917 when a lead smelter was opened for processing and refining ore. As the operation grew, the mine processing area grew to 6,200 acres (2,509 hectares) with mines that were one mile (1.6 km) deep with 125 miles (201.2 km) of mine tunnels at peak production. It was one of the largest mining districts in the world. The nearby Coeur d'Alene River provided a water supply for ore processing, agriculture, and drinking and was convenient for transportation of ore, supplies, and refined metals which promoted the growth of the facility.

The processing of lead at Bunker Hill involved crushing of the mined ore. Then waste rock was removed by hand and/or mechanical sorting. The lead ore was mixed with limestone flux and heated to just below the melting point to cook off water and oxygen. This cooked ore was then heated in a blast furnace where carbon in the flux removed oxygen and sulfur to float at the top of the vessel and allowed the molten lead to sink to the bottom where it could be siphoned off for casting. The nonmetal by-products that rose to the top of the molten lead were skimmed off and discarded, forming slag waste.

Enormous quantities of hazardous waste were produced in these processes. Much of the mine tailings, wastewater, and waste dust were dumped untreated into the Coeur d'Alene River. The slag was dumped around the site which produced contaminated runoff when it rained and the wind blew and spread contaminated dust most of the year. This toxic release continued throughout the 100+ years of mining. The lead-bearing dust contaminated the soil and washed into surface and groundwater in the vicinity of the mine, smelter, and mill. The local streets were sanded with mine and smelter tailings in the winter and it was used as backfill in construction projects. This dispersion of lead produced a serious threat to the health of the Bunker Hill workers and the residents of nearby towns in Idaho and eastern Washington. Lead in the Coeur d'Alene River basin caused the deaths of large numbers of fish, invertebrates, and waterfowl.[41]

In response to the dangerous environmental conditions at the facility in the early 1900s, a baghouse was installed to filter exhaust and dust from lead refining at the smelter in 1923.[42] This filter still released a lot of lead dust to the environment for fifty years or more. In 1938, a tailings pond was built from low-grade ore and waste rock in the floodplain of the Coeur d'Alene River. The pond received and stored process and mine water. The lead-rich particulate settled in the pond prior to the water being discharged into the Coeur d'Alene River. However, the pond only retained some of the lead wastes from Bunker Hill and the rest was discharged.

The mining and ore processing at the Bunker Hill complex expanded through the 1940s and 1950s. Because the mine operation provided such high-paying jobs, the workers and residents did not complain about the poor environmental conditions. By the 1970s, the Bunker Hill smelter had the second-highest lead production in the United States. This was 20 percent of the lead production in the world and more than 20 percent of the lead and zinc production in the United States. Between 1965 and 1981, the smelter released about 6 million pounds (2.7 million kg) of lead to the atmosphere. In 1973, the baghouse was damaged in a fire which increased lead emissions to the environment and it was not repaired for more than a year and a half.

However, in the 1970s, the Federal Clean Air Act and Clean Water Act regulations forced the company to make significant environmental improvements. They constructed a wastewater treatment plant to treat the water from the tailings pond before being released into the Coeur d'Alene River. They repaired and improved the baghouse by 1975. However, the Bunker Hill company fought all of the required changes and purchased and destroyed all of the houses within a half mile (0.8 km) of the smelter rather than allow officials to examine and test them. They also replaced contaminated soil with fresh topsoil in many locations. However, their actions did not stop the Idaho Department of Health and Welfare from testing lead in air, soil, and vegetation in the area.[43] They determined that there were excessive lead concentrations throughout the area.

The company attempted to contest these appalling test results by commissioning their own research study. They collected and tested local children's urine samples without obtaining the parents' consent. The company never released these results. However, mounting public pressure forced them to fund and cosponsor a major program to test lead levels in local children's blood by an independent research team. In January 1975, the results were released and 170 of the 172 children tested who lived near the smelter had blood lead levels above 40 μg/dL. For example, the children in the town of Kellogg averaged 50 μg/dL of blood.[44] The Centers for Disease Control (CDC) regards 5 μg/dL as cause for concern and children with levels above 45 μg/dL are recommended for treatment. They further found that 22.1 percent of the 172 children had lead levels above 80 μg/dL in their blood. The CDC recommends emergency medical treatment for anyone with blood lead levels above 70 μg/dL. There was even one child who had a blood lead level of 160 μg/dL.[45] This was the highest blood lead concentration ever recorded in the United States.[46]

The mine and smelter closed and 2,000 people lost their jobs because of the rampant pollution and costs to address it. In 1983, the EPA designated a twenty-one-square-mile (54.4 km^2) area as the Bunker Hill Mining and Metallurgical Complex Superfund site. This is one of the largest Superfund sites in the United States.[47] The cleanup of the area involved treatment or removal of tens of thousands to millions of tons of tailings, waste rock, and lead-contaminated soils from numerous sites. Lead-contaminated soil was removed from sixteen public parks, as well as 2,000 private and commercial properties. The mining company declared bankruptcy in 1994 as a result of the staggering remedial and legal costs. At the smelter, a 32-acre (13-ha) mine waste landfill was built, closed, and capped. More than 1.25 million cubic yards (1 million m^3) of lead contaminated soil were disposed of on-site. More than 30,000 cubic yards (22,937 m^3) of contaminated sediment was removed from the Coeur d'Alene River. In 2002, the EPA expanded the Superfund site to include the 1,500-square-mile (3,885 km^2) Coeur d'Alene Basin and declared it as one of the largest and most polluted areas in the United States. By 2016, Lake Coeur d'Alene was still loaded with lead and it was killing wildlife. Local residents were still suffering mental and physical disabilities.

FLINT, MICHIGAN, LEAD POISONING

A less severe but more infamous case of lead poisoning occurred in Flint, Michigan, in 2015.[48] City water was contaminated with lead exposing many residents to lead and causing a public health crisis. It resulted in several high-level city, state, and even federal officials being terminated and prosecuted. Flint has been deemed by many as a case of environmental racism because of the high number of African American residents affected.[49]

The city of Flint, Michigan, is the original home of General Motors Company which began in 1908. The city was a boomtown as a result with high employment rates and vast wealth. Many of the public water lines were installed between 1901 and 1920. The water mains are cast iron but the lines installed from the mains to and within homes were lead. By the 1950s and 1960s, Flint's population grew to nearly 200,000, as people flocked to careers in the booming automotive industry.[50]

This massive industrialization badly polluted the Flint River which flows through the center of the city. Factories along the river dumped treated and untreated waste into the river water and dumped garbage along the banks. These industries included lumber and paper mills, carriage and car factories, and meatpacking plants among others. Raw sewage from the city's waste treatment plant was piped into the river and agricultural and urban runoff and leachate from landfills added to the pollution. The massive waste in the Flint River caught fire twice during this time and had to be extinguished.

In the 1970s–1980s, however, high oil prices and increasing sales of imported cars resulted in autoworker layoffs and plant closures throughout the area. In response, Flint lost its wealthy citizens and the population decreased by half. The city turned into a poverty and crime-ridden slum with one in six houses abandoned. The municipal budget plummeted and most public projects had to be abandoned. All replacement projects of the 43,000 water lines including 3,500 lead lines, 9,000 galvanized steel lines, and 9,000 lines of unknown type was halted. In 1967, Flint began receiving its public water supply through the Detroit Water and Sewerage Department (DWSD) from Lake Huron with the Flint River as an emergency backup water source. This DWSD supply system ended in 2013 because the city of Flint experienced a catastrophic fiscal crisis and could no longer afford it. The city switched water supplies to save about $5 million over two years.

The new water supply system was initiated on April 25, 2014. The city developed a construction plan for a new water line from Lake Huron at this time that would take about thirty months to complete. However, DWSD contested the plan which slowed the construction project. In the meantime, the city obtained its drinking water from the Flint River as a temporary solution.

The residents of Flint immediately began complaining about the color and smell of the tap water.[51] The city applied high amounts of chlorine to the Flint River water supply to reduce bacteria but it increased the trihalomethanes levels and they are carcinogenic. The city realized they had a problem by late summer but did not release this information until January 21, 2015. Unfortunately, the chlorine also did not adequately control bacteria in the drinking water as an outbreak of Legionnaire's disease began in June and the river water was suspected. This outbreak lasted at least through November 2014 causing eighty-seven infections and twelve deaths. The outbreak was not announced until 2016. Also, a water boil advisory was issued from August 14–20 and again in September.[52]

The excess chlorine in the municipal water started corroding the engine parts at Flint's General Motors Truck Assembly plant and, as a result, they stopped using it in October 2014.[53] Even this setback did not convince the city to address their water problems. However, on February 26, the EPA tested a lead level at seven times the regulation limit in one Flint home.[54] The reason that the lead was so high was not only because of the excessive chlorine but also that city did not treat the Flint River water with corrosion inhibitors as was done in the past. As a result, the corrosive tap water was leaching lead out of the old lead pipes as well as iron which colored the tap water yellow to bright red. The Flint council members were so alarmed by this finding that they immediately voted to reconnect to DWSD on March 23 but the decision was vetoed by the city emergency manager. A well-publicized battle between city officials and scientists with the public over the safety of Flint's water ensued.

The EPA found extremely high lead levels in the tap water in four Flint homes on June 24.[55] In response, on July 9, the mayor of Flint drank tap water on television to prove it was safe. In addition, Michigan Department of Environmental Quality (MDEQ) officials reported that the Flint water quality was safe and potable on a radio show on July 13. This stance had to be hastily reversed, however, when on September 8, an extensive research study was published reporting that 40 percent of Flint homes had elevated lead levels in their tap water. Soon after, on September 24, a Hurley Medical Center pediatrician reported that Flint children were experiencing elevated blood lead levels since the supply was switched.[56]

This tsunami of evidence turned public opinion sharply against the city government. The governor of Michigan signed a bill on October 15 to reconnect Flint to DWSD for $9.35 million that was enacted the following day. The mayor of Flint declared a state of emergency on December 15 and the Michigan National Guard began distributing bottled water on January 12, 2016. The governor made a request to President Obama to declare Flint as a federal disaster area. He agreed and made the declaration on January 16, directing $5 million in aid to the city.

Lead leached from aging pipes into household water exposed all 100,000 Flint residents to dangerous levels of lead including 6,000 to 12,000 children. The mishandling of the water crisis compelled four city, state, and federal officials to resign or to be fired. Then in January 2021, the then former governor of Michigan and eight senior officials were charged with thirty-four felonies and seven misdemeanors including two officials being charged with involuntary manslaughter.[57]

THE TAKEAWAY

These legacy lead pollution incidents are thankfully rare but by no means have they all been uncovered. Others will continue to be reported in the media but in decreasing regularity as they are remediated. The reason that the incidents are so shocking at this point is that ambient lead in the environment, in general, has been so thoroughly reduced that the contrast makes them appear immense. In reality, much of the population was exposed to these levels of lead on a daily basis. It was only through the directed efforts of Pat Patterson and his supporters and advocates and resulting public pressure that forced the government against their will to enact legislation to regulate environmental lead. The scientific research, proposed solutions, public education, and public pressure all acted in the proper sequence and amount in the right areas to resolve this terrible crisis. It is a noteworthy example of the victory of humans over a serious public health threat against the wishes of industry and the government. This is an outstanding example of how the combination of science, an eminent scientist, and public pressure can help society overcome a very serious environmental and public health crisis.

Chapter 4

Fixing a Hole in the Ozone

Perhaps the most serious environmental threat of the 1980s and 1990s was the developing hole in the Earth's protective stratospheric ozone layer. This threat was caused by human activity and it resulted in a worldwide panic before action was taken. The resolution of the hole in the ozone layer is probably the ideal example of how environmental issues should be handled and this type of response should be the goal for addressing climate change and any other environmental threat in the future. The problem of degradation of ozone by a pollutant was first hypothesized to be possible by research scientists. Then, the resulting hole in the ozone layer was found by a scientific research team. Its existence was widely publicized in the media causing great concern by the general public who pressured elected officials to act to address it against industry wishes. A worldwide treaty was developed and signed by most nations and action is underway to resolve the problem by about 2050 to 2070. And this solution is working.

WHAT IS OZONE?

Ozone is a highly reactive form of oxygen that instead of the usual configuration of O_2, it is the compound O_3. Ground-level ozone is hazardous to human health and makes up some of the most dangerous parts of air pollution. It forms on hot days in the summer when primary air pollutants of nitrogen oxide, largely from automobile exhausts, reacts in the atmosphere with volatile organic compounds (VOC) from evaporating gasoline, paint thinner, acetone, and other organic solvents.[1] Both of these are considered primary pollutants because they come directly from an emitting source. The hot sun drives the photochemical reaction between the two pollutants in the atmosphere producing the secondary air pollutant of ozone. It is secondary because it is a reaction among primary pollutants. Ozone is common in cities in the summer but especially in hot cities with poor circulation.

Ozone-rich pollution is sometimes called Los Angeles smog because it is so common there.

Ground-level ozone is sometimes called "bad ozone" because of the detrimental health effects it can cause. Exposure causes coughing, sore throat, labored breathing, inflamed airways, increased lung infections, worsening of lung diseases like asthma, emphysema, and chronic bronchitis and an increase in asthma attacks.[2] It is especially dangerous to people with asthma, young children, older adults, and people who spend a lot of time outdoors. It is estimated that more than 6,000 people die each year from exposure to ozone but it shortens the life span of as many as 1 million people annually.

In contrast to this bad ozone at ground level, stratospheric ozone is "good ozone." It is naturally formed in the lower stratosphere as the result of incoming ultraviolet light or radiation from the sun.[3] The energy from this radiation causes some O_2 molecules to break apart into simple O which then reacts with other O_2 molecules to form O_3. The electrical discharge from lightning can also form ozone from a similar splitting and recombining process. Ground-level ozone is common around electrical generation plants and transfer stations for the same reason.

This stratospheric ozone is not very stable. The ultraviolet rays that split the O_2 molecules also split O_3 molecules into O and O_2. In addition, ozone is dissociated into O and O_2 by chemical reaction with nitric oxide (NO), which can occur naturally in the atmosphere.[4] Besides the devastating impact of some explosive volcanic eruptions that injected gas and ash into the stratosphere, the processes that create and destroy stratospheric ozone operated in relative equilibrium prior to the impacts of humans. The concentration of ozone in the stratosphere was directly modulated by the amount of ultraviolet radiation from the sun. This stratospheric ozone and its production and dissociation reactions occur in the ozone layer, a zone approximately thirteen miles (21 km) thick that completely envelops the earth in the lower stratosphere. Even here, the ozone is very sparse with only about three ozone molecules for every ten million molecules of air.

BENEFITS OF THE OZONE LAYER

The ozone layer shields the Earth from about 98 percent of the dangerous ultraviolet radiation from the sun.[5] The most dangerous, longer-wavelength ultraviolet (UV) radiation is directly absorbed by the ozone layer. If this radiation reaches the earth's surface, it causes serious negative health impacts on humans and most other terrestrial life on Earth. It causes eye damage including cataracts, premature aging, and immune response suppression. The most serious health impact from exposure, however, is an increase in skin cancer in humans. The UV radiation damages DNA in skin cells to the point where the damaged and mutated cells grow quickly and multiply and ultimately, skin cancer develops. There are two main types of skin cancer, non-melanoma skin cancer, including basal-cell cancer and squamous-cell cancer, and melanoma.[6] Approximately 90 percent of the non-melanoma skin cancer in the United States is the result of exposure to ultraviolet radiation from the sun. The incidence of this cancer increased sharply by 77 percent from 1994 to 2014. On a worldwide basis, more than 64,800 people die of non-melanoma skin cancer annually. In terms of melanoma, at least 99,780 cases of the more dangerous, invasive type of melanoma were reported in 2022 in the United States causing about 7,650 fatalities. Between 2012 and 2022, the cases of invasive melanoma increased annually by 31 percent. Clearly, skin cancer is a very dangerous disease.[7]

DESTRUCTION OF THE OZONE LAYER

The threat to the ozone layer from human activity was first recognized by Paul Crutzen, one of the most influential environmental scientists in the world.[8] Crutzen was from the Netherlands and suffered terrible ordeals at the hands of the Nazis, barely surviving. Although he began a career as an engineer, he later switched to atmospheric research and, as the result of participation with a scientist from the United States, he began investigating stratospheric ozone distribution. This interest and research led him to complete a PhD at Stockholm University in 1968, on photochemical theories for stratospheric ozone. He completed a postdoctoral research fellowship on stratospheric ozone at Oxford University, United Kingdom from 1969 to 1971 but then returned to Stockholm University to earn a Doctor of Science degree in 1973. His dissertation research was titled "On the photochemistry of ozone in the stratosphere and troposphere and pollution of the stratosphere by high flying aircraft," which was groundbreaking at the time. This work and its extension that he pursued in subsequent years earned him the 1995 Nobel Prize in chemistry, which he shared with Mario Molina and F. Sherwood Rowland.

The greatest achievement of Paul Crutzen's career was his pioneering work on possible processes for the depletion of the ozone layer.[9] He first discovered that nitrogen oxides could cause ozone depletion in 1970 while at Oxford University. He proposed that these nitrogen oxides, resulting from internal combustion engines and heavy application of agricultural fertilizer, could rise upward into the stratosphere. There, a series of chemical reactions driven by ultraviolet radiation could dissociate the ozone into O and O_2. This hypothesis became even more concerning in the late 1970s, when supersonic transport airplanes became popular and were planned to replace conventional jet airplanes. These high-speed aircraft fly at higher altitudes than normal jet airplanes and release nitrogen oxide-rich exhaust at levels where they would cause great damage to the ozone layer. However, market and economic factors precluded the building and use of large numbers of these aircraft which may have helped preserve the ozone layer, at least temporarily. Therefore, although valid, this mechanism wound up not being the main culprit in damaging the ozone layer.

ROLE OF REFRIGERANTS

This part of the ozone layer story involves the technology of refrigeration. In an effort to preserve food longer and ensure more stable food supplies, a system of refrigeration was required to cool the food and therefore slow the decay.[10] The early efforts of refrigeration mostly utilized ice boxes. To accomplish this, ice was harvested from northern lakes in the winter and stored in icehouses. The ice was then transported throughout and even outside the country and delivered by icemen to residences and businesses. The global ice trade was centered in the northeastern United States and in Norway, at least, at first. It formally began in 1806 and peaked in the mid-nineteenth century though it was practiced locally prior to this. The industry expanded around the world and transitioned from natural lakes to artificial impoundments and to mechanical-chemical refrigeration before contracting by the early twentieth century and eventually terminating.

Although mechanical-chemical refrigeration was first discovered in 1755 and improved throughout the nineteenth century, ice was only progressively replaced by chemical refrigeration in home refrigerators starting in 1913. The basis of refrigeration is heating a volatile liquid chemical in a sealed system where it is compressed and pumped into an expansion chamber. The change from a liquid to a gas cools the entire system by latent heat transfer. The early refrigerators were dangerous because they used hazardous volatile chemicals like ammonia, methyl formate, methyl chloride, or sulfur dioxide and they commonly leaked. As a result, refrigeration units had to be located outside the house or have an elaborate system that kept the chemicals away from people. Even then, there were still multiple cases of poisoning and deaths from them.

Everything changed with the invention of a wonder compound by a research team in 1928 called Freon which was first patented by Frigidaire on December 31, 1928.[11] Freon is one of a class of chemicals called chlorofluorocarbons (CFCs) that were originally developed in the 1890s but not widely applied until this renewed interest took hold. Freon is chemically nonreactive, nontoxic, nonflammable, and noncorrosive so the dangers of previous refrigerants were avoided. In addition, it is very stable so it does not break down or require periodic replacement as a result. The boiling point of Freon is about room temperature and it easily transitions from a gas to a liquid, which makes it an excellent refrigerant for everyday uses. The inventor was so confident in the stability of Freon that he inhaled a lungful of it at a public chemistry convention to prove its safety. To pursue this invention, the General Motors and DuPont companies collaborated to form the Kinetic Chemicals Company in 1930 to produce dichlorodifluoromethane which is known as "Freon-12," "R-12," or "CFC-12."

Freon revolutionized the refrigeration industry. Over the next five years, more than eight million new refrigerators were sold in the United States using Freon as their coolant.[12] In 1932, the Carrier Engineering Corporation developed the first self-contained home air-conditioning unit using Freon. It was also later used as a propellant for aerosols for cosmetics, pharmaceuticals, insecticides, paints, adhesives, and cleaners among others. Additional CFC compounds were developed including trichlorofluoromethane (Freon 11), chlorodifluoromethane (Freon 22), trichlorotrifluoroethane (Freon 113), dichlorotetrafluoroethane (Freon 114), and chloropentafluoroethane (Freon 115) and each had a number of differing uses. For example, Freon 11 was used as a refrigerant, a foaming agent in polyurethane foams, a solvent and degreaser, and in fire extinguishing chemicals. Freon 113 was used in the manufacture of fire extinguisher chemicals, in drying and degreasing of electronic parts and equipment, and as a dry-cleaning solvent for fabrics, leather, and suede in addition to a refrigerant. This meant that Freon was widely released to the atmosphere both by leakage from cooling and refrigeration but also deliberate release in aerosol applications. Freon compounds developed into among the highest volume chemical groups manufactured in the United States at the time.

The cooling of homes and buildings resulting from the development of Freon allowed large population shifts to southern areas of the United States that previously were underdeveloped because of the excessive heat.[13] It allowed work to be conducted in comfort in most industries, improving productivity. It also allowed the development of many new industries. It seemed like a miraculous invention but it had a dark side that would soon become apparent.

PREDICTED PROBLEMS WITH CFCS
THROUGH RESEARCH

There are very few cases when a research scientist discovers a potential major global problem, it is verified, and the world acts to address it, but this is exactly what happened with Freon. Mario Molina was the scientist who conducted the scientific research.[14] He was born in 1943, in Mexico City, Mexico, and chose to pursue physical chemistry in 1968 when he entered the PhD program at the University of California at Berkeley. He completed a PhD in 1972 and a postdoctoral fellowship the following year. In 1973, Molina completed a second postdoctoral fellowship at the University of California at Irvine and his advisor was Dr. F. Sherwood Rowland. It was there that he began his groundbreaking work on CFCs.

Molina began research on the purely scholarly topic of CFCs and their interactions with atmospheric gases while at University of California, Irvine. It was an open area of study because CFCs are relatively inert in the lower atmosphere. Therefore, the other atmospheric chemists did not expect any significant findings so chose not to pursue it. Even Molina had no reason to expect that these inert CFC compounds could be a problem. However, he found that solar radiation could dissociate CFCs at high altitudes, which produced highly reactive free chlorine atoms. Molina and his advisor, F. Sherwood Rowland, determined that these free chlorine radicals were chemically destroying ozone in the stratosphere. They published this theory in the highly prestigious journal *Nature* in 1974.[15]

The process is possible because CFCs are so stable and long-lived, they are able to be swept to altitudes of 15 to 25 miles (24 to 40 km) above the Earth's surface without chemically breaking down.[16] This altitude is well into the stratosphere where ultraviolet radiation can dissociate them just like it does to ozone and normal oxygen. This dissociation dislodges free chlorine atoms from the CFCs which are highly reactive with many other compounds. The free chlorine (Cl) can then participate in the ozone destroying reactions and yielding O_2 and ClO from the ozone. In turn, the new ClO reacts with free oxygen (O) to reform free chlorine and O_2. The newly free chlorine then attacks another ozone molecule and breaks it down as well. This cycle of dissociation and reforming reactions can persist for as long as two years and destroy as many as 100,000 ozone molecules for every CFC molecule in the stratosphere. Further, the ClO-producing reaction consumes the free oxygen ions preventing them from combining with O_2 to produce more ozone. In this way, it also slows ozone production.[17] Once the 100,000 cycles are complete, the chlorine is typically removed from the stratosphere by reacting to form hydrogen chloride or chlorine nitrate which sink into the troposphere and are washed to the surface by precipitation.

Drs. Molina and Rowland explained this process in their groundbreaking paper where they also calculated that if CFCs continued to be released at the rate of that time, the Earth's ozone layer would be drastically depleted and thinned in just a few decades. They further showed that approximately 85 percent of the chlorine in the stratosphere came from dissociation of CFCs. For this pioneering research on ozone chemistry, Molina and Rowland shared the 1995 Nobel Prize in chemistry with Paul Crutzen but it was not without controversy.

THE BATTLE TO BAN CFCS

CFCs were widely used industrial chemicals and part of a huge production industry. In 1976, the United States produced more than 374,786 tons (340 million kg) of CFCs.[18] The CFC industry employed 200,000 people and it was valued at over $8 billion (US). It produced a slew of refrigeration, cooling, and aerosol products that were used in most aspects of everyday life in industrialized countries. They would not admit to the problem nor stop producing products without a serious battle. However, this was not the usual battle between the leader of the research findings and an industry. Molina was young and interested in teaching at a university, not controversy. In 1975, he obtained the position of assistant professor at University of California at Irvine. In order to obtain tenure and keep their job, assistant professors must publish, obtain grants, and establish a good reputation in the scientific field. Fighting a social battle would guarantee that Molina would make enemies and jeopardize his career. Instead, Rowland assumed the task of trying to make the world appreciate the importance of their research and its implications even though stratospheric ozone chemistry was not his major area of expertise.

Rowland gave speeches at every opportunity to promote a worldwide ban of CFCs and, as a result, was attacked professionally and personally.[19] Du Pont dismissed the theory as "purely speculative" and refused to otherwise acknowledge it. They reasoned that there was no proof that the ozone layer was being damaged. More derogatory statements like it was "a science fiction tale," "a load of rubbish," and "utter nonsense" were also repeated across the industry and to the press. Rowland even had the theory accused of being orchestrated by the USSR Ministry of Disinformation of the KGB. It was not just industry either. Rowland lobbied the press, state legislators, congressmen, senators, and anyone with influence to take action against CFCs to even more ridicule. However, his lobbying did lead to nine congressional hearings that he and New Mexico Senator Pete Dominici held on ozone depletion in 1975 and 1976. Even at some of these hearings, Rowland was subjected to personal and professional attacks. Unprofessional elected officials called him a "maverick" and an "oddball" who was "off the deep end" and a "publicity seeker."

Perhaps even more embarrassing, Rowland was also attacked by his professional colleagues. The Molina and Rowland paper did not contain experimental evidence but instead, focused on the hypothesis that CFCs caused ozone depletion based purely on other supporting hypotheses. If just one of these supporting hypotheses was proven to be false, it could undermine the entire hypothesis. Some noted research climatologists attacked the methodologies, instrumentation, and calculations used in the paper on a professional level.

The pressure on and criticism of Rowland were greatly relieved in 1976 when the US National Academy of Science released two reports titled "Halocarbons: Effects on Stratospheric Ozone"[20] and "Halocarbons: Environmental Effects of Chlorofluoromethane Release."[21] Other scientific studies were soon released that also supported Molina and Rowland. Several studies even proposed that they had underestimated the risk to the ozone layer. However, even these supportive studies did not slow Rowland's effort but his detractors diminished. For more than a decade, he continuously urged the banning of CFCs. Unfortunately, there was no evidence that the ozone layer was being damaged for a long time after it was predicted.

THE HOLE IN THE OZONE LAYER

The actual identification of the hole in the ozone layer was almost as strange as the rest of the situation. It began with the hiring of a young analyst, Jonathan Shanklin, by the British Antarctic Survey in 1984.[22] He was asked to analyze years of data on stratospheric ozone density from the Dobson ozone spectrophotometer at the Halley Research Station in Antarctica. The instrument determines the amount and type of UV radiation reaching Earth. It is also therefore an indirect measure of the amount of ozone in the atmosphere above Antarctica. These data had not been properly analyzed for years and important information on them was simply scrawled on numerous sheets of paper. When the data were compiled and properly analyzed, the scientists were shocked to learn that the stratospheric ozone was greatly reduced over the South Pole. In fact, since the late 1970s, there had been a steady decline in the spring ozone concentration to the point that by 1984, the ozone layer in Antarctica had thinned by about one third. They published these findings in the journal *Nature* to the shock of the world and vindication of Molina and Rowland.

The entire complexion of the issue changed with this discovery. All doubt of the CFC-ozone connection evaporated and research on stratospheric ozone took off.[23] It was determined that the highest depletion occurs over Antarctica with lesser depletion over the Arctic. The reason that ozone over the Antarctic is so sensitive to CFCs is because of the geography. Antarctica is mountainous and surrounded by open ocean which causes the ozone layer temperatures to drop below −108°F (−78°C) in the winter. This temperature exacerbates ozone depletion because more clouds form above the area which help drive the chemical reactions that release free chlorine atoms from the CFCs. The freed chlorine then breaks down ozone during the following spring. Even Molina returned to his ozone research and developed a model to explain the CFC-ozone processes.

The public response to the news was equally impressive. The discovery of the hole in the ozone layer spurred massive publicity and educational campaigns by scientists and environmental activists aimed at the public and law makers. Even though the mid-latitudes where most people live suffered no loss of ozone, there was widespread public alarm and even panic, at least locally. The onslaught to ban CFCs was so great that industry offered little resistance to it. CFCs were quickly banned from aerosol propellants and other applications and legislation was initiated to ban them from all applications, including refrigerants, on a nearly global basis. The world spectacularly united to defeat this global environmental crisis.

As a result, the United Nations quickly brokered the Montreal Protocol on September 1, 1987, an international treaty that required developed countries to stop using CFCs and banning all uses of the most aggressive CFCs by 1996.[24] This protocol was unanimously supported and is widely regarded as one of the most successful international environmental agreements ever. It was initiated in 1989 and regulates production and consumption of about 100 chemicals designated as ozone depleting substances (ODS). The treaty requires reductions in consumption and production of the ODS in a stepwise manner, with unique timetables for each ODS as well as for developed versus developing countries. ODSs include CFCs, carbon tetrachloride, methyl chloroform, hydrochlorofluorocarbons, methyl bromide, and hydrofluorocarbons.[25]

Hydrochlorofluorocarbons (HCFCs) are used in refrigeration, air-conditioning, and foam applications, and include Freon/refrigerants R-22, R-123, R-124, and R-142b. HCFCs are also greenhouse gases with the most common HCFC having 2,000 times more global warming potential than carbon dioxide. Because of the climate impact, it was decided to accelerate the HCFC phaseout in September 2007. Developed countries were to complete the HCFC phaseout by 2020. Developing countries started phasing out HCFCs in 2013 and will complete it by 2030. Hydrofluorocarbons (HFCs) are non-ozone depleting alternatives to CFCs and HCFCs that were developed and implemented to support the phaseout timetable. They are being used in air conditioners, refrigerators, aerosols, and foams among other products in place of CFCs and HCFCs in all new products. Unfortunately, HFCs have global warming potentials of 12 to 14,000 times greater than carbon dioxide. Their emissions have been increasing at 8 percent per year and will rise to 7 to 19 percent of global CO_2 output by 2050. The Montreal Protocol parties agreed to reduce usage of HFCs on October 15, 2016. They agreed to a gradual reduction of 80–85 percent by 2050 and chemical replacements to be developed. These reductions began in 2019.

The Antarctic ozone hole size during the designated benchmark month of September accelerated quickly from nonexistent in 1978 to about 7.7 million square miles (20 million km²) by 1987.[26] The size increased slowly after that, peaking between 1998 and 2007 at about 10.4 million square miles (27 million km²). It then slowly decreased to an average of about 8.9 million square miles (23 million km²) between 2010 and 2021. Therefore, the Antarctic ozone hole is healing very slowly.[27] The problem is that CFCs can have atmospheric lifetimes as long as fifty years. As a result, even in the absence of continued emissions stratospheric ozone will not completely recover until after 2070.

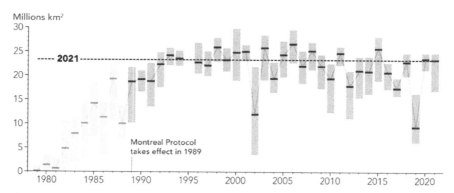

Figure 4.1.   Graph of the annual size/area of the hole in stratospheric ozone layer over the Antarctic in September–October from 1979 to 2022. The black line is the average and the gray box is the variation. Note the quick growth in the 1980s, peak between about 1997 and 2008, and the gradual reduction afterward.
Source: EPA.

THE TAKEAWAY

The global response to the CFC damage to the stratospheric ozone layer has been environmentally responsible. The identification of the potential impact of CFCs on the ozone layer was prior to there being a problem. Although there was some industry pushback initially, there was relatively quick acceptance of the research findings by the profession. The identification of the damage to the ozone layer was therefore not unexpected and society responded appropriately.[28] This sequence of tests and responses should be the model for the resolution of climate change and all other environmental problems in the future. Besides the resistance encountered by Rowland, all parties responded appropriately and the problem is being resolved.

Chapter 5

The Air That We Breathe

Even before the industrial age, air pollution was causing public health crises if not disasters on an occasional to regular basis. Once the industrial age began, these incidents got much worse and in the early to mid-twentieth century they became regular disasters.[1] After some widely publicized disasters with significant loss of human life caused by unrestricted industrial emissions, the public became very concerned to outraged and took action. They pressured the governments to enact legislation to restrict industry from releasing toxic emissions at will. Restriction on air pollution was the first broad and serious environmental legislation enacted. The components of acid precipitation, sulfur and nitrogen compounds, were and are two of the most dangerous air pollutants. As a result, they were targeted in legislation and underwent near miraculous reductions to the point of preindustrial levels in most ambient air in the United States. This was a very serious public health and environmental problem but it was resolved nearly completely, at least locally, in a very short period of time by using the right approach.

Although humans have polluted oceans, surface water, groundwater, soil, the upper atmosphere, and even outer space, it is the fouling of the air that has caused the most concern. It was the first form of pollution to compel serious environmental legislative action to be taken in industrial times. This enhanced concern is likely because the air envelops us and all humans must inhale it every few seconds to stay alive. Polluted air cannot be ignored like many other polluted media because it is never out of sight. It is also common that people have an immediate negative health reaction to polluted air like wheezing, coughing, and having their eyes tear. Also, the most immediate and devastating pollution disasters have been through the air. The stories about the experiences of the people and horrifying photographs of the victims are quickly posted by the media spreading fear worldwide. Further, air pollution disasters tend to be in cities with large populations so they impact large numbers of people. In many cases, it takes these public health disasters and dangerous situations for people to take action.

NATURAL AIR POLLUTION CRISES

Unlike many of the other media, even natural pollution of air can be cata-strophic and humans have had to deal with it forever.[2] The biggest single source of natural air pollution is a volcanic eruption. These events can deliver huge amounts of gaseous chemicals and particulate to the troposphere and even the stratosphere. Summit eruptions can shoot gas and ejecta more than 15.5 miles (25 km) into the air and expel as much as 672 cubic miles (2,800 km[3]) of lava, gas, and ejecta into the air. The gases can contain large enough amounts of sulfur and fluorine compounds to poison the biota within hundreds of miles of the volcano. They can deliver enough ash to irritate and damage human and animal respiratory systems for thousands of miles. The aerosols they release can disperse incoming sunlight causing the earth to cool significantly for several years after the eruption. Volcanic pollutants can cause the deaths of millions of people and immeasurable numbers of plants and animals.

The main chemical pollutants from volcanoes are sulfur compounds. The 1912 eruption of Novarupta in Alaska was the largest of the twentieth cen-tury, emitting 3.4 cubic miles (14 km[3]) of ash and sulfur compounds which produce sulfuric acid when mixed with water.[3] Novarupta began erupting on June 6 and it continued for sixty consecutive hours, plunging an area hun-dreds of miles around the volcano into complete darkness because of the thick ash. The sulfur compounds poisoned local streams, making them acidic and devastating the salmon populations. It took many years for them to recover. It is said that the ash cloud swept hundreds of miles southeast through the state of Washington and when it contacted wet clothes drying on clotheslines, it was so acidic that it burned holes in them.

Probably the most polluting volcanic eruption in recent history was the 1783 Laki eruption in Iceland.[4] From June 8, 1783, to February 7, 1784, Laki produced 3.5 cubic miles (14.7 km[3]) of lava, the largest outpouring in historical times. It also produced excessive amounts of fluorine and sulfur compound vapors which settled over Iceland and produced a dry fog known as the "Laki haze" for five months. Fluorine combined with water to produce deadly hydrofluoric acid and sulfur emissions produced sulfuric acid the fallout from which poisoned grazing areas, killing 80 percent of the sheep in Iceland, 50 percent of the cattle, and 50 percent of the horses. By 1785, 20–25 percent of the human population (>10,000 people) had also died from poisoning and starvation. The Laki haze spread eastward across Europe and the acid fallout devastated vegetation and human health there as well.[5] Crops outright failed in Iceland but had low yields or even failed throughout Europe that summer. There were fish kills in Scotland from acid deposition. Between the multiple famines and health impacts of the Laki haze, it is estimated that as many as six million people died worldwide. The disruption in the food supply led to widespread poverty and unrest through 1788 and is considered a major contributing factor in the French Revolution in 1789.

Sandstorms and haboobs can sweep enormous amounts of sand and dust aloft where they can travel thousands of miles impacting as many as millions of people along the way.[6] The fine particles are not typically toxic but inhaling them can damage the respiratory system and exacerbate asthma, bronchitis, and COPD. Dust from the Sahara Desert is regularly swept across the Atlantic Ocean and impacts people in the Caribbean and the southern United States. Extensive worldwide loess deposits were formed by fallout of airborne dust and sand and document that the process has been occurring throughout geologic history.

Wildfires also emit huge amounts of particulate matter into the atmosphere as well as some dangerous chemicals.[7] Smoke from wildfires in Colorado has been known to darken skies all the way to the East Coast of the United States. It too can cause severe respiratory issues. Although wildfires can be caused by human carelessness or arson, they can also be caused by lightning and be natural. Radon is a naturally occurring air pollutant from the natural decay of uranium in rocks and soil. Radon in outdoor air is relatively safe but it is radioactive and can be concentrated enough in indoor air to be deadly.[8] Exposure to high levels of indoor radon can cause lung cancer. In fact, radon is estimated to cause as many as 25,000 deaths from lung cancer per year in the United States alone making it the most dangerous natural environmental pollutant. Even asbestos is a naturally occurring mineral. It is infamous for causing the lung cancer mesothelioma from exposure in industrial settings but the original source is in rocks.[9] Just under 3,000 people die in the United States per year from mesothelioma. There are many other sources of natural air pollutants from salt spray to pollen and spores but they tend to be less dangerous.

ANTHROPOGENIC AIR POLLUTION

Humans have always had to endure this natural air pollution and besides not using asbestos or building structures that exclude radon, there is not much that can be done about it. Humans have also contributed extensively to air pollution. The first pollution came from using fire and that was the full extent of the human contribution for tens of thousands of years. However, once humans began to process chemicals, the pollution greatly increased in both amount and toxicity. Initially, this pollution was primarily from processing mineral ores into metals which releases extensive sulfur compounds and many other chemicals into the air but also from any grinding of minerals or plants which releases dust or particulate into the air. The use of flammable pitch, oils, and alcohol contributed dangerous volatile organic compounds (VOCs) from evaporation of these liquids and polycyclic aromatic hydrocarbons (PAHs) from burning them.[10] These usages were uncommon and very localized so did not contribute much to large scale air pollution. The development of gunpowder significantly added to it, at least locally like in battles.

The real air pollution problems began during the industrial revolution or just preceding it. The use of coal for heating, cooking, and manufacturing greatly increased air pollution especially in urban areas.[11] Coal emits a number of dangerous compounds when burned. It contains sulfur compounds so it produces acid fallout and precipitation but also carbon monoxide, PAHs, mercury, and especially soot. Soot is particulate which is any very small solid or liquid particles that are suspended in air. It is categorized by size with total particulate matter (TPM) being all suspended matter between 100 and 0.1 microns.[12] Within this group, PM10 particulate is particles less than 10 microns and PM 2.5 is less than 2.5 microns. PM10 particles can be inhaled but sit on the outside of the skin in the throat, lungs, and sinuses. PM 2.5 particles can be absorbed directly into the bloodstream through the lungs making them very dangerous especially for people with certain medical conditions. Primary particles are PM10 or larger and emitted directly from the sources as solids whereas secondary particles are PM 2.5 and form in the atmosphere through photochemical reactions. Much of fine particulate is from chemical reactions among sulfate chemicals from power plants, nitrates from vehicle exhausts and industrial emissions, and ammonia from animal feed lots. Particulate of 50 to 100 microns settles close to the source or is washed out of the air by precipitation but all particulate can be transported great distances in suspension depending upon the wind velocity and consistency.

Greater than 99 percent of inhaled particulate is exhaled or trapped in the upper respiratory tract by mucus and later expelled.[13] Coarse particulate can embed in lung tissue and cause shortness of breath including life-threatening asthma and COPD. Fine particulate absorbed into the bloodstream through the lungs is especially dangerous to people with heart or pulmonary disease. Children and the elderly are also more sensitive to particulate. Short-term exposure may cause coughing, wheezing, irritation, and labored breathing but long-term exposure to particulate increases instances of lung diseases such as emphysema, pulmonary fibrosis, and lung cancer. It also damages the immune system. Chronic exposure to particulate can shorten a person's life by two years. It is estimated to cause up to 65,000 deaths in the United States per year and as many as 200,000 deaths in Europe per year.

Particulate and chemical pollutants have the potential to cause public health disasters under certain weather conditions. The worst of these conditions is a temperature inversion which can seal concentrating air pollutants at ground level rather than allowing them to disperse as they do under normal conditions. In this case, a lid of cold air seals the warm, continuously polluting air underneath it. This can be caused by topography where mountains or highlands surrounding a city in a basin can block air circulation like in Los Angeles, California, or Denver, Colorado. However, it can also occur in cities without topographic barriers under certain weather conditions. Several impressive air pollution disasters have occurred through this mechanism, both by topography and weather.

1930 Meuse Valley Disaster

The first recognized major air pollution disaster in the industrial age occurred in early December 1930 in the Meuse River Valley of Belgium.[14] The Meuse Valley is an industrial area that was populated with coke ovens, blast furnaces, zinc smelters, fertilizer factories, and sulfuric acid plants at the time. The 9,000 residents worked both farms and in the industries. The twelve miles (19.3 km) between Liege and Huy was the most industrial stretch of the valley at this time but it is also very narrow and bordered by cliffs rising 270 to 350 feet (82.3–106.7 m) on both sides. This topography caused almost constant air pollution incidents because of the topographic temperature inversions it led to. Most people considered the air pollution to be an unavoidable side effect of the industrial prosperity, and little pressure was applied to the offending industries to control it as they released enormous quantities of particulate, sulfur dioxides, and other gases. In 1911, many cattle died as a result of increased air pollution events. In the town of Engis, windows were etched by repeated releases of hydrofluoric acid.

A topographic temperature inversion between December 1 and December 5, 1930, in the Meuse Valley, resulted in the first reported major industrial air pollution disaster. At the time, it was calm and cool in the Meuse Valley with no wind and high barometric pressure. Above the valley walls, however, dense, humid air formed a thermal blanket over the valley, resulting in a thick fog that prevented polluted air from moving out and it quickly built up. By December 3, hundreds of people were suffering from laryngeal irritation, uncontrollable coughing, chest pain, and labored breathing. Others had cyanosis, asthma, nausea, vomiting, and pulmonary edema. By December 5, sixty-three people were dead from respiratory failure and they were primarily children, the elderly, and people with medical conditions. On December 5, the weather changed, and the public health crisis quickly passed.

The main irritant in this disaster was the acidic chemical, sulfur dioxide (SO_2), at concentrations of at least five times the current safe levels. The SO_2 was produced by the burning of coal in the valley, which was concentrated in the air due to the temperature inversion. The disaster drew widespread attention of the public and even governments all across Europe.

1948 Donora, Pennsylvania, Disaster

The next major air pollution disaster was in Donora, Pennsylvania, in 1948 and it attracted public attention all across the United States as well.[15] Donora was an industrial community, forty miles (64.5 km) southwest of Pittsburgh with a population of about 14,000 in the 1940s. Donora had the two largest metal working mills in the United States at the time. The four-mile (6.4-km) long series of industrial complexes was in a horseshoe bend of the Monongahela River surrounded by limestone cliffs rising 500 feet (152.4 m) above the valley floor. The nearby coal mines and the convenience of transporting materials and products by barge on the river made this an ideal industrial area. Donora claimed to be the largest maker of nails in the world at the time. The coal-fired furnaces for iron and zinc production expelled sulfur oxides, zinc, fluoride, other metals, coal ash, and soot, all discharged through 150-foot-high (45.7 m) smokestacks, which released these emissions well below the cliff heights.

Air pollution in the area steadily increased over the years. From 1918, the zinc company agreed to compensate residents for health care costs related to pollutants released by the plant. Since the 1920s, the mill owners were forced to pay residents for pollution damage to crops and livestock. However, from early morning on October 26, 1948, and for the next four days, a true air pollution disaster occurred in Donora as a result of the trapping and accumulation of these industrial emissions. On October 25, a blanket of cool, dry air settled over the town, forming a topographic temperature inversion against the cliffs. The sulfur oxides, metals, and particulates from the inappropriately low smokestacks quickly built to dangerous levels in town. The air turned yellow and then gray. By Friday, October 29, the town was covered in a dark haze, making it difficult to see. The next morning, nine elderly residents of Donora were dead from asphyxiation and by the following day, it was eighteen. Doctors urged people with respiratory problems to evacuate, but the thick fog and resulting congested streets made it impossible. By Sunday, the town funeral home ran out of caskets and florists ran out of flowers as 20 people between 52 and 85 were dead. The local hospital was mobbed with very sick people. At least 7,000 residents were suffering with headaches, nausea, and vomiting. Firefighters went door to door giving residents a few gulps of oxygen. The inversion then abruptly lifted and the pollution abated.

However, fifty additional residents died of respiratory complications over the following month from having been exposed to the air pollution. Sulfur dioxide in air was estimated to have reached between 19 and 69 times the current legal limits.

The Donora disaster was reported in all news media nationwide and the public became very concerned. It resulted in Pennsylvania enacting the first air pollution control law in the United States in 1955. This disaster among several others led to US Congressional hearings on air quality in the United States and, ultimately, the first Clean Air Act was signed into law in 1963.

1952 Great London Air Pollution Disaster

The worst urban air pollution disaster ever occurred just four years later in London, United Kingdom.[16] In the 1950s in London, coal was commonly used to heat homes and to cook, and it was heavily used by industry throughout the country. Coal smoke was an additional component of the frequent fogs in London leading to the term "smog" as a combination of the two. If coal is burned at low temperature and without enough air, it gives off excessive smoke. This contains hot gases of carbon monoxide, sulfur compounds, and even mercury but also small carbonaceous particles that produce soot. The place that most typified the need for control of particulate air pollution was the city of London, which has historically been filled with smoke and smog and covered with soot. This particulate is made of microscopic particles of less than fifty micrometers (μm) and is a recognized health hazard. Soot was a problem in most industrialized cities but especially in London.

On December 5, 1952, a layer of cold air over London trapped warm, moist air in an unusually long-lasting temperature inversion. Pollutants emitted by industry, coal-burning stoves, and furnaces began to rapidly accumulate in the fog, dropping visibility to just a few feet. The next morning, residents of the city woke to darkness. The pollution quickly began to cause serious health problems in the at-risk populations who ventured outside and the city hospitals were soon overrun with patients suffering from respiratory issues. Chaos in the streets and poor visibility forced ambulances to stop running, leaving thousands of suffering people to walk through the smog to the hospitals. At least 500 people died that day from respiratory complications.

By December 7, the visibility in London was reduced to one foot (30 cm), making it as dark as night throughout the day. Roads were blocked by abandoned cars making it impossible to evacuate anyone. Smog infiltrated homes and buildings causing health problems even for people sheltering indoors. A theater was closed because smog inside the building was so thick, it was difficult for patrons and performers to see or even breathe. At least 750 victims perished from inhaling the pollution that day. On December 8 and 9, air quality and public health conditions continued to deteriorate causing more than 900 deaths each day. Smog was so penetrative that it was reported to be present deep in the archive rooms of libraries. Animals in markets were reported to be dropping over dead from the pollution. Road, rail, and air travel were essentially impossible, isolating the city from the outside world. Fortunately, by the end of December 9, the weather changed and the inversion lifted quickly, clearing away the pollution. However, the deaths caused by exposure to the smog continued for several weeks. Deaths from bronchitis and pneumonia were calculated at more than sevenfold normal rates as a result of inhaling the smog.

Health officials initially estimated that the Great London smog disaster caused the deaths of about 4,000 people. Later studies estimated, based on the increased mortality through the month of December 1952, that the smog disaster actually caused about 12,000 deaths. The average sulfur dioxide is estimated to have been well in excess of twenty times the current legislated maximum and total particulate concentrations were greater than six times the maximum.

Through public pressure, the British Parliament eventually admitted to the severity of the disaster and, with even stronger public pressure, passed the 1956 Clean Air Act of the United Kingdom. This legislation exerted tighter controls on air pollutants and required stricter monitoring standards. In spite of this new legislation, in 1956 another killer smog struck London, killing an additional 1,000 people with the same causes and results. However, the legislation led to the reduction of coal usage and better filtering technologies for smokestacks and other exhausts.

LOS ANGELES AIR POLLUTION BATTLE

Although the occurrence of deadly, localized, inversion-type air pollution disasters generally decreased in intensity through the 1960s, the total amount of air pollution greatly increased. The increase in automotive traffic increased the amount of nitrogen oxides, sulfur oxides, lead, particulate, deadly polycyclic aromatic hydrocarbons (PAHs), and volatile organic compounds (VOCs) in ambient air. There was also extensive atmospheric nuclear device testing through the 1950s and until 1963, when an international treaty ended it. This testing injected massive amounts radioactive material into the atmosphere. Fallout of this material to the surface was worldwide, nearly continuous for several years and impacted essentially everyone. There were ever-increasing industry emissions and new chemicals being developed and released, causing air pollution to intensify across the industrial world and beyond. Asbestos was being used in numerous industrial settings but even the general public was exposed to it in everyday life, in many cases.

These pollutants progressively degraded urban air quality in industrialized countries but it even degraded the overall quality of the air on a global basis. One of the worst places for urban air pollution was Los Angeles, California.[17] Even though Los Angeles developed an extensive urban mass transit system which peaked in the early 1940s, it steadily declined as automobiles, inexpensive gasoline, and the freeway system blossomed. As a result, the rail service was discontinued in 1961, and streetcar use ended in 1963. It took thirty years for another Los Angeles city rail system to be constructed. The increase in automobile, bus, and truck usage, the increased electrical generation, refinery production, and general burning of fossil fuels concurrently increased during the 1940s through 1960s. The combination of an increase in air pollutants in the Los Angeles basin with the high temperatures, surrounded by high sheltering mountain ranges, and moderated by the cool Pacific Ocean, all contributed to make the air essentially unbreathable most of the time.

The configuration of these surrounding mountain ranges led to the formation of regular topographic temperature inversions that trapped air pollutants in the city. Cool ocean water lowered the temperature of the air near the surface while warm, overlying air formed a lid that kept the cooler air from mixing or dispersing. The cool air flowed over the city from the west and trapped pollutants, causing early morning to midday fog. By 1943, this fog had already become polluted. The Los Angeles Bureau of Air Pollution Control was formed in 1944 to investigate the building air pollution but this and other efforts were ineffective. By 1946, Los Angeles air pollution had grown much worse, leading to the formation of the Los Angeles County Air Pollution Control District (APCD) in 1947. By 1951, the APCD achieved reductions in smoke from iron foundries and mills, reduced the evaporation of hydrocarbons, and later controlled sulfur dioxide emissions. With all industrial sources of air pollution in Los Angeles regulated, air quality should have improved but it did not. Attention began focusing on the ever-increasing automobile exhaust.

It was discovered that nitrogen dioxide from automobile exhaust in the presence of evaporated gasoline (volatile organic compounds or VOCs) and exposed to ultraviolet light produces ozone which is deadly. Ozone-rich smog is generally referred to as L.A.-type or photochemical smog. This smog was causing significant adverse health effects in the urban residents. It was not welcome news. By 1954, there were 2.4 million motor vehicles in Los Angeles, the most of any city in the world and they consumed 5 million gallons (18.9 million L) of fuel per day. In 1966, APCD estimated that

16,000 tons (14,515 mt) of air pollutants were produced daily in Los Angeles and 80 to 90 percent of it came from automobiles.

In 1960, the California Motor Vehicle Pollution Control Act was enacted and regulated a limit of 275 parts per million (ppm) for hydrocarbons and 0.5 percent for carbon monoxide in ambient air. By August, at least 100 companies were developing anti-smog devices for California automobiles to attain this limit. After June 1965, every new car sold in California had a device and the major American automobile manufacturers developed their own emission controls. The California Air Resources Board (CARB) replaced the APCD to better enforce the new laws. The 1968 California Pure Air Act was passed and required state car fleets to use hybrid, steam, and natural gas vehicles. Despite all of these regulations, it took well into the 1980s for the pollution in Los Angeles to begin to abate.

Los Angeles was not the only city with dangerous air pollution problems.[18] For example, New York City also had serious air pollution including a temperature inversion from January 15 to 24, 1953, which caused pollutants to reach high levels. Sulfur dioxide reached nearly 450 times normal levels in the city. At least 162 people died as a result of this pollution disaster. Another air pollution inversion incident struck New York from November 27 to December 4, 1962. It caused about forty-five deaths. There were three other events between 1962 to 1964 which caused air quality to become an issue in the concurrent mayoral race.

CAUSE OF ACID PRECIPITATION

One of the most common and dangerous pollutants in each of these disasters is sulfur dioxide (SO_2). Along with nitrogen oxides (NOx), it produces acid precipitation and was deadly in the disasters described and in the 1960s, in general. Industrial sources of SO_2 are primarily from the burning and refining of fossil fuels.[19] In the past, electrical-generating power plants accounted for 67 percent of SO_2 emitted in the United States. Coal-burning plants produced about 96 percent of these emissions. About 18 percent was from industrial and residential fuel combustion including car emissions and home heating. The rest was from other fuel usage in non-road vehicles including ships, trains, construction equipment, as well as metal processing and petroleum refining.

The health impacts from exposure to SO_2 result from its reaction with water to form sulfuric acid and other related acids.[20] Short-term exposure to it irritates the nasal passages, throat, eyes, and lungs and causes coughing, and shortness of breath. Higher dosage during a moderate pollution event causes headache, dizziness, nausea, vomiting, and skin and eye burns. High exposure, like in a pollution disaster, causes convulsions, pulmonary edema, and death. Symptoms appear at 5 parts per million (ppm) within 5 to 15 minutes of exposure, 20–50 ppm exposure, causes immediate skin and eye burns, and less than 10 minutes of exposure to 1,000 ppm causes death. Long-term exposure to SO_2 like in chronically polluted cities causes permanent lung damage such as chronic bronchitis, and other respiratory illness, inhibition of thyroid function, decreased fertility, and premature death. It is particularly dangerous to those with asthma, lung disease, and heart disease, and children and the elderly. Even low doses could cause life-threatening reactions in these people. There is a direct and proven correlation between high SO_2 levels in major cities and increases in hospital admissions.

The other very dangerous air pollutant that contributes to acid precipitation is oxides of nitrogen or NOx. Oxides of nitrogen include nitric oxide (NO), nitrogen dioxide (NO_2), and nitrous oxide (N_2O), among others. NO_2 and N_2O are directly toxic to humans and animals and NO is damaging to them. Natural sources of NOx include soil bacteria, volcanic eruptions, forest fires, and lightning. Human sources of NOx are from motor vehicles, power generation and utilities, and industrial, residential, and commercial sources. Commercial sources include welding, electroplating, engraving, dynamite, chemical production, dyes, lacquers, rocket fuel, and TNT. Large amounts of NOx are emitted to the atmosphere from nitrogen-based fertilizers. NOx is also in tobacco smoke.

Direct exposure to NOx is dangerous to human and animal health.[21] NOx combines with water to produce nitric acid both in the air and in the respiratory system. Low-level, acute exposure to NOx in the air like that in everyday air in many American cities in the 1960s irritates eyes, nose, throat, and lungs and results in coughing, shortness of breath, fatigue, nausea, and vomiting. Within a day or two of higher-level exposure, fluid can build up in the lungs and may be accompanied by anxiety, confusion, lethargy, and fainting. High levels of NOx like in an air pollution disaster can cause burning, spasms, and swelling of the upper respiratory tract resulting in asphyxiation. Fluid filling the lungs can be fatal. Exposure to high levels of NOx in the air can also seriously burn the skin and eyes possibly leading to blindness. Long-term exposure like in chronically polluted cities can cause permanent damage to the lungs, heart, and nervous system.

In addition to the air pollution disasters, all of which included excessive amounts of SO_2 and NOx, the residents of many urban areas in industrialized nations suffered from the health effects of exposure on a regular basis by the 1950s and into the 1970s. It was not uncommon to see photographs of residents of Tokyo, Japan, wearing surgical masks on a regular basis to protect themselves from acidic air pollution, fallout (dry deposition), and precipitation.[22] The acidity reached such high levels in many urban and industrial areas, including several cities in Japan, that just getting caught in a rainstorm could result in the need for medical attention. Eye and skin damage and increased respiratory problems were common in many of these cities as a result of the acid precipitation and acid fog or mist. Many older residents as well as those with respiratory and pulmonary diseases limited their trips outdoors for fear of suffering serious medical reactions. The main culprits in this danger were SO_2, NOx, ozone that results from reactions involving NOx, and particulate or soot.

Another problem resulting from acid precipitation is the enhanced dissolution and deterioration of human-made structures. This was of special concern for ancient structures. Antiquities in many urban areas, especially those made of marble, were being eroded and dissolved by the acid rain at an alarming rate. In several ancient cities in countries like Italy and Greece, traffic was restricted to keep it away from the more famous and at-risk structures such as the Coliseum in Rome and Parthenon in Athens. However, it was not only the ancient structures that were at risk. The acidity dissolved all marble facades but also cement in concrete and unprotected iron. This is especially damaging to bridges and trestles which are surrounded on all sides by polluted air and are regularly flexed by weight changes from the passing traffic and trains. This flexing can flake off paints and coatings and contribute to cracking and spalling of concrete and cement. Acid precipitation figures prominently in the degradation of infrastructure.

This urban air pollution even spread outside of urban areas.[23] Acid emissions from coal-burning power plants and other industrial sources in Michigan and other Midwestern states fueled acid rain all the way to natural areas in the Adirondack Mountains of New York. Because there is no natural buffering of surface water acidity from rocks in the higher Adirondacks, 90 percent of the lakes there became acidic and devoid of aquatic life during the 1970s. In addition to the Adirondacks, acid rain caused acid surface waters throughout the northern Midwest, the Middle Atlantic states, and parts of the Southeast among others. Native aquatic species cannot survive in acidic surface water. In Virginia, 6 percent of the streams could not support trout, and 50 percent of the other streams were badly impacted. In Pennsylvania, the surface water conditions were worse. Acid precipitation also damaged trees at higher elevation from Maine to Georgia, reducing growth and reproduction and making them susceptible to disease and insects. Even lower-elevation vegetation, including agricultural crops, were adversely damaged by the acid fallout and precipitation. It also stripped nutrients from open soil reducing both forest and farmland productivity.

Sulfur emissions have long been a problem for human civilization. This is because many of the metals used by humans are from ores that are rich in sulfur and smelting releases it to the environment. This has been occurring for thousands of years. However, during the industrial revolution, sulfur emissions increased dramatically as a result of the ever-increasing burning of sulfur-rich fuels, especially coal, and the equally increasing demand for metals.[24] Enhanced emissions began about 1870 in most industrialized countries but accelerated the fastest in the United States. By 1920, sulfur emissions were already excessive, having increased by 10- to 15-fold. Sulfur emissions in Europe also increased during this period but at a much slower rate of three- to fourfold and the rest of the world was minor. This situation changed radically starting between 1940 and 1950 and world emissions of sulfur had tripled by 1970. Of these, emissions in Europe increased by about 2.5-fold during this time but East and Central Asia increased by about five-fold over the same period. The USSR was a big contributor to this increase. The United States increased as well by about 50 percent.

Nitrogen emissions had a similar trend to sulfur, beginning from about 1870 and accelerating in the United States until the late 1920s with very little output from other countries. The increase was much slower than SO_2 but it is not as easily calculated. Also, similar to sulfur, NO_2 emissions increased dramatically between the late 1940s and 1970. The global increase during this time essentially quadrupled NOx emissions mostly led by Europe, East and Central Asia (especially Japan, Taiwan, and Korea), and the Soviet Union. The emissions in the United States increased as well but not as radically as the other countries and areas, partly because it was already elevated.

LEGISLATION AND REDUCTION OF SO_2 AND NOX

Most of the air pollution disasters and marked deterioration of air quality in many urban areas in the United States, Japan, and Europe occurred during this period of radically increasing SO_2 and NOx emissions from the late 1940s to 1970.[25] Also during this time, some minor air pollution laws were passed but none of them alleviated the situation. The big change came when the US Congress passed the Clean Air Act of 1970. Later that year, the US Environmental Protection Agency or EPA was established to enforce it. Great Britain established an equivalent environmental agency in the same year. The US 1970 Clean Air Act named six "criteria" air pollutants including both SO_2 and NOx as well as carbon monoxide, lead, ozone, and particulate matter as the most dangerous and in need of reduction. It required the EPA to establish National Ambient Air Quality Standards (NAAQS) or exposure limits for these six criteria pollutants.[26] They developed research-based quantitative standards to protect public health and the environment. The pollutants were effectively targeted in all industry, federal, and municipal sources. This was the most powerful piece of legislation to clean the air to date and it was quickly followed by similar legislation in many other industrialized nations.

The US Congress directly took on the problem of acid precipitation by passing the Acid Deposition Act in 1980.[27] This law initiated an eighteen-year research and assessment effort through the National Acidic Precipitation Assessment Program (NAPAP). This program expanded a network of monitors to determine the details of acidic precipitation in order to determine long-term trends and establish a network of at least 280 stations nationwide to measure dry acid deposition. NAPAP quantified the effects of acid precipitation on a regional basis to determine the effects on freshwater and terrestrial ecosystems under different conditions. In addition, NAPAP assessed the effects of acid precipitation on building materials, historical buildings, and monuments. The program funded extensive studies on the atmospheric processes and its role in controlling acid precipitation.

In 1990, Clean Air Act Amendments were added to the 1970 Clean Air Act.[28] They included more stringent air-quality standards and motor vehicle emission controls. They required control of acid rain, study and plans for the use of alternative (non-hydrocarbon) fuels, elimination of some toxic air pollutants, and the phasing out of stratospheric-ozone-destroying chemicals. They also required the Environmental Protection Agency to revise the National Ambient Air Quality Standards for the six criteria pollutants. They developed quantitative primary standards to protect public health and secondary standards to protect the environment and infrastructure. The problem is that the new 1990 standards rendered many areas that had worked hard to meet the 1970 standards now in noncompliance. It also regulated 188 other hazardous chemicals that had not been regulated prior to this.

There was also international regulation on NOx after 1970. In 1997, the Kyoto Protocol of the 1992 United Nations Framework Convention on Climate Change (UNFCCC) was ratified by thirty-seven countries but not the United States. The protocol commits participants to reduce greenhouse gas emissions including NOx. It went into effect in 2005 with the first commitment period from 2008 to 2012 and the second period from 2012 to 2020 with separate goals for each. There was also an earlier agreement—the 1988 Sofia Protocol was enacted to control emissions of NOx. In this protocol, there were twenty-five participating countries that agreed to reduce NOx emissions to 1987 levels. The United States volunteered to reduce NOx emissions to 1978 levels in this protocol.

The SO_2 reduction methods that were applied fall into three groups: (1) use of low sulfur fuels, (2) removal of sulfur from fuel prior to combustion or fuel desulfurization, and (3) removal of sulfur from the waste gases after fuel combustion, known as flue gas desulfurization (FGD).[29] The most common method of SO_2 reduction from smokestack emissions is with a device called a scrubber. The EPA committed to reduce SO_2 emissions by a massive 10 million tons (9 million mT) between 1980 and 2010. As part of this program, they targeted coal-fired power plants in two phases, first in 1995 and second in 2000. Phase one addressed the power plants with the highest emissions and phase two addressed smaller plants. The older power plants were grandfathered in by the EPA during phase one and, in response, were used so heavily that sulfur emissions actually increased from 1992 to 1998, but phase two restricted their use and emissions markedly decreased. Another potent regulation was the requirement for ultralow sulfur diesel fuel (<15 ppm sulfur) in all cars, trucks, buses, and other vehicles since 2006.

A real boost to the reduction of both SO_2 and NOx was the increased use of natural gas to generate electricity.[30] This occurred as a result of increased availability of natural gas through the enhancement of drilling and fracking technology beginning in about 2005. The advancement was the new ability of drillers to drill horizontally. Prior to this, wells had to be drilled near vertical so a horizontal target rock layer (which most are) could only produce gas from the thickness of the layer. By drilling to the producing layer and then turning the well horizontal, much more of the layer could be encountered and produced. Companies began drilling shale layers that contain a lot of oil and gas but have very low permeability. In these cases, the oil and gas cannot flow to the well for production. Fracking is the process of pumping a water-sand mixture into the well and shale rock at very high pressure until the rock cracks. The sand is driven up into the cracks where it holds them open so the oil and gas can flow more easily to the well. Many petroleum-bearing areas of the United States and around the world that had previously been too expensive to produce now became targets. The great increase in inexpensive gas availability encouraged the construction of many new gas-fired power plants and the abandonment of the older coal-fired power plants. Gas burns much cleaner than coal and has much less sulfur. This conversion greatly helped with SO_2 emission reductions.

These efforts collectively resulted in a precipitous decrease in SO_2 but it took several decades.[31] Estimates are that at least 31.2 million tons (28.4 million mT) of sulfur dioxide were emitted into the air in the United States in 1970. By 1980, total emissions had decreased by a respectable 17 percent but by 1990, it was only down by 24 percent overall indicating a slowing of reduction. However, by the mid to late 1990s, when new restrictions went into effect, sulfur emissions reductions ranged between 37 and 41 percent from original showing a renewed drop. It decreased again to about a 52 percent reduction from the 1970 benchmark by 2001 but stayed about the same through 2006 when fracked natural gas began to be used. Then the SO_2 emissions fell dramatically to about an 84 percent reduction from 1970 in 2012. In 2021, just 1.8 million tons (1.6 million mT) of sulfur dioxide were emitted in the United States, a phenomenal 94 percent reduction since 1970. The efforts had been successful and the threat was averted.

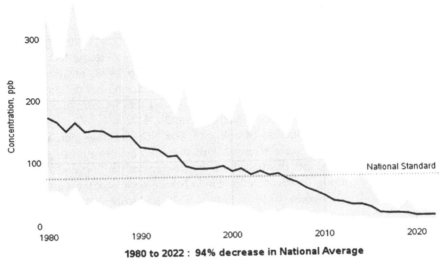

1980 to 2022 : 94% decrease in National Average

Figure 5.1. Graph of SO$_2$ content in ambient air in the United States from 1980 to 2022. The solid line is the average and the gray area is the range across all monitoring stations. The National Standard is the public safety level above which action must be taken. The United States crossed that line in 2005 and all monitoring stations have been below it since 2011.
Source: EPA.

Perhaps even more impressive was the reduction of sulfur dioxide concentration in the air that people and animals breathe. According to the EPA, ambient air had an average of 170 parts per billion (ppb) in 1980 which was already reduced from the levels in 1970. In comparison, for the period from 2017 to 2021, the average concentration was 20 ppb, an 88 percent reduction from 1980 levels and a 94 percent reduction overall. Most of the reduction was since 2000 when the concentration was still 110 ppb. This means that there was a 53 percent reduction since 2000. The national standard for sulfur dioxide is 75 ppb so the air we breathe on a daily basis is nearly sulfur-free. It is estimated that current air is at about the same sulfur dioxide concentration as it was in 1890.[32] This, however, only applies to the United States, Europe, Canada, and a few other countries. The sulfur dioxide emissions in China increased 27 percent during the time that those of other countries were being reduced. Fortunately, the decreasing sulfur dioxide emission efforts are outweighing the increases and global levels peaked in about 1980 and have decreased significantly ever since.

The technologies to control NOx release fall into two categories: (1) controls on the sources of NOx and (2) controls on emissions of NOx from combustion.[33] Controls on the sources of NOx reduce the amount of NOx formed in the combustion areas of a furnace as fuel is burned. These technologies are combustion controls and low–NOx burners (LNBs). Controls on post-combustion NOx emission reduce NOx concentration in the exhaust gas as it passes out of the smokestack, tailpipe, vent, or chimney. These technologies include both selective and nonselective catalytic reduction (catalytic converters), conventional reburning systems, and fuel-lean gas reburn (FLGR) systems.

In 1970, NOx emissions in the United States were about 27.5 million tons (25 million mT). Whereas five of the six criteria air pollutants had decreased significantly by 1998, NOx had not. In fact, it had increased by 10 percent by some estimates but it had certainly not decreased as markedly as the others.[34] In response, efforts to reduce NOx were redoubled. One of the main NOx emissions abatement technologies was the addition of catalytic converters to automobile exhaust systems. The main reduction from this technology began in 1995. Oxygenated fuels, hybrid engines, compressed natural gas (CNG), and fuel cells also reduced these emissions. With this reduction, by 2006, 25 percent of NOx emissions were from coal-fired power plants. As a result, the Department of Energy required them to use new clean burn technologies. However, this became less important with the increase in use of fracked natural gas. Clean-burning natural gas has low nitrogen content and not much NOx is generated from burning. In comparison, higher nitrogen fuels, such as coal and oil, produce significant NOx in the exhaust. Even these fuels can produce lower NOx exhaust by lowering the flame temperature. This is done by modifying the burner to a low flame combustion system called low NOx burners, or LNBs. These systems can reduce NOx generation by as much as 30 to 50 percent.

As a result of all of these efforts, NOx emissions began to decrease precipitously.[35] Estimated total NOx emissions dropped from more than 24 million tons (21.8 million mT) in 2002 to about 7.5 million tons (6.8 million mT) in 2021. The reduction of NOx from all of the efforts is about 73 percent in the United States. Most other industrialized countries in Europe and elsewhere experienced similar reductions. However, unlike SO_2, NOx continues to increase on a global basis.[36] Global atmospheric N_2O averaged between 270 and 275 ppb until about 1890 when it increased slowly to just above 285 ppb by 1950. Since then, it skyrocketed to about 335 ppb by 2010.

THE TAKEAWAY

Although the deadly air pollution disasters have apparently been resolved, there are cities and areas in the world where air pollution remains as a great threat to public health and the environment. Acid precipitation still exists in

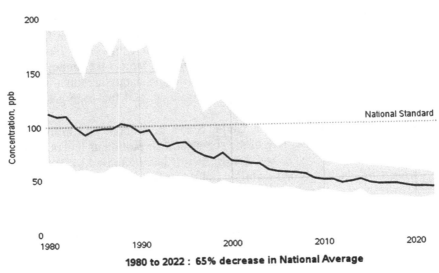

1980 to 2022 : 65% decrease in National Average

Figure 5.2. Graph of NO_2 content in ambient air in the United States from 1980 to 2022. The solid line is the average and the gray area is the range across all monitoring stations. The National Standard is the public safety level above which action must be taken. The United States crossed that line in 1998 and all monitoring stations have been below it since 2003.

Source: EPA.

some locations as well. However, in many industrialized areas where danger-ous sulfur and nitrogen compounds were targeted for action, they were nearly miraculously reduced. Considering how dangerous these pollutants are to human health, surface water, vegetation, and wildlife, and how widespread they were, their resolution is a major environmental human victory for many industrialized nations. Having SO_2 reduced by 94 percent in the United States means the problem has been completely resolved. Clearly, the countries that have not addressed the problem could reduce or resolve them as well, follow-ing the same or similar course of action. People and governments have the ability to work together and overcome even severe and widespread problems like air pollutants. The acid rain resolution provides an excellent template for resolving the climate crisis and any other environmental problem.

Chapter 6

Supertanker Superaccidents

A short-duration threat but one that caused local devastation to the environment was the poor planning for transporting large quantities of crude oil in supertankers. When these huge ships were damaged and leaked phenomenal amounts of oil in ecologically sensitive coastal areas, besides the great expense, they caused essentially unrecoverable environmental damage or at least damage that would last for a number of decades. After several spectacular accidents that were well covered in the media raised the outrage of the general public, laws were passed to require the tankers to be built and operated following certain specifications that protected them from breakage under most conditions. This is not to say that oil tanker accidents that spill excessive amounts of oil do not still occur but after being an epidemic, they are now rare.

CRUDE OIL

Although fossil fuels are now considered evil by some environmental activists, crude oil is a remarkable substance.[1] As marine microorganisms die and settle to the seafloor, they become part of muddy sediments which are later compressed into shale rock as subsequent sediments added on top compress and heat them. The organic part of the rock from the microorganisms transforms first into kerogens and later into crude oil and natural gas with increasing pressure and temperature. The oil and gas migrate from this source rock to geological features where they can be accumulated called traps. It is these deep underground traps that are drilled and produced to provide oil and its many uses in society.

Crude oil is a complex mix of many chemicals with numerous important uses, not just as fuel. Almost all plastics are made from petroleum as are paints, coatings, industrial solvents, components of medicines, fertilizers, pesticides, and most industrial production chemicals. Even if new sources and systems of energy are developed, society will still need to mine crude oil for the many needed chemicals. Actually, burning petroleum as an energy source is a very poor use of this valuable substance.

These chemicals are primarily separated from each other in an oil refinery in a distillation tower.[2] The crude oil is heated at the base of the tower or even outside the tower until it vaporizes. The released vapor rises inside the tower and separates into the different chemicals which individually rise to a height dictated by their density. They are condensed at these levels through cooling and collected for use. For example, road tar condenses at the lowest level generally followed by diesel fuel, kerosene, gasoline, and propane at the top. These basic products can be used as is or engineered into needed products. These petroleum products are primarily long-chained hydrocarbons that can be broken into shorter chains as needed using catalysts in a process called "cat cracking." The refinery engineers the petroleum products into whatever public demand dictates at the time. For example, heating fuel is engineered and produced in the winter but not in the summer.

Even each of these basic products contains numerous chemicals.[3] For example, gasoline contains the "BTEX" compounds including benzene, toluene, ethylbenzene, and xylene which are all highly volatile and very toxic. When gasoline is exposed to air, these chemicals evaporate quickly and pollute the air as volatile organic compounds or VOCs. Before they evaporate, however, they are toxic enough to kill most wildlife from large organisms to microbes. It is for this reason that spilled crude oil is deadliest when it is first released. If it is spilled in the open ocean, it spreads out into a thin layer that is heated by the sun and well exposed to the air. It evaporates quickly and does minimum damage to wildlife because large fish and marine mammals can swim away and it cannot easily reach the seafloor. In coastal areas or on land, however, it is deadly to all wildlife and capable of causing ecological disasters especially if it is spilled in high quantities.

TRANSPORTATION OF CRUDE OIL

The first oil well was drilled in Titusville, Pennsylvania, in 1859, and drilling sprouted up all over the country and much of the world after the uses of oil were discovered. Once it became clear that petroleum was the best fuel for most transportation, heavy machinery, and many industrial processes, demand skyrocketed. Oil is inexpensive, easily produced, transported, and stored and the energy output cannot be exceeded per unit volume by any other fossil or biologic fuel. Its convenience and affordability in personal automobiles and other vehicles propelled car sales and resulting demand set new records every year especially in the 1950s and 1960s. America developed a love affair with cars as a result and oil was an integral part.

The problem is that petroleum is not available everywhere. Certain regions produce huge amounts of oil whereas most areas do not. The Gulf of Mexico area, the Persian Gulf region, and several areas of the former Soviet Union are just a few examples of areas capable of producing many billions of barrels of oil every year. The ever-increasing demand for gasoline quickly exceeded the supply of oil from local smaller deposits. It forced companies to construct large, centralized oil refineries to process the crude oil into gasoline. This excessive demand and usage and centralization of production and refining placed inordinate pressure on companies to develop efficient methods to transport both crude oil and refined gasoline and other petroleum products. On land, freight trains and tanker trucks could handle most of the distribution of gasoline and refined products but getting the crude oil from the oilfield to the refinery across an ocean presented an inordinate challenge.

The solution to this problem was the development of tanker ships which could transport large amounts of crude oil from the fields across the oceans and to the refineries.[4] Tankers to transport oil were already in use by 1900. However, with increasing demand more and larger tankers had to be built. The demand expanded greatly during World War II but almost solely for the war effort. The postwar period in the 1950s saw an explosion of domestic demand for gasoline with many Americans buying cars and driving everywhere on the new interstate highway system. By the 1960s, the expanding demand led to a desperate need for larger oil tankers, and the shipbuilding industry responded. Before 1961, all oil tankers were less than 100 long tons in deadweight tonnage. The first larger tankers were built in 1961 with a dry weight of more than 100 long tons but less than 150 long tons. Very soon after, in 1967, the first tankers with weights greater than 200 long tons deadweight tonnage were constructed. These were the first supertankers, and they were built in ever increasing numbers after that. In a short twenty-year period, oil tankers had increased in volume by tenfold.

SUPERTANKER ACCIDENTS

It was bad enough for the environment when a smaller tanker had an accidental release of oil in a near-shore area. Collisions could release a huge amount of crude oil all at once in a small area. Because this crude oil was still in a fresh state, it still contained all of the volatile toxic compounds and therefore exposed all local wildlife to it.[5] This exposure killed or impaired all marine organisms and terrestrial birds, animals, invertebrates, and microbes that could not flee quickly enough. Unfortunately, most tanker accidents were in coastal regions causing ecological disasters. The spillage of large amounts of crude oil and resulting ecological disasters became one of the major issues in the American environmental movement. A big event in this interest was the Santa Barbara oil spill from an oil well blowout on January 28, 1969. It released about 80,000 to 100,000 barrels of oil at 42 gallons per barrel. It wreaked havoc on the local wildlife and was the top story in newspapers and on the evening television news for almost two weeks, infuriating the American public. Accidental spills from supertankers fueled this outrage.

Torrey Canyon Disaster

It did not take long once supertankers were being constructed for a terrible accident and ecological disaster to occur. The *Torrey Canyon* supertanker was involved in an accident in 1967.[6] Although it was first built in 1959 at 60,000 tons, it was later enlarged to 120,000 tons (108,862 mT), making it the thirteenth largest ship in commercial use at the time. The ship was 974.4 feet (297.0 m) long, 125.4 feet (38.2 m) wide, and had 68.7 feet (20.9 m) of draught. Its collision and spill is still the United Kingdom's worst environmental disaster and it was the first oil tanker accident to attract extensive worldwide media coverage.

The American-owned but Liberian-flagged supertanker *Torrey Canyon* was fully loaded with 119,328 tons (108,252 mT) of crude oil in Kuwait and began its final voyage on February 19, 1967. The ship maneuvered south around the Cape of Good Hope, up the west coast of Africa past the Canary Islands and arrived at the Isles of Scilly off the southern English coast by March 17. The tanker was bound for the BP oil refinery in Milford Haven, southeastern Wales. The Scilly islands are a tourist attraction and a vacation spot for many people from Britain and western Europe. There are at least fifty-five islands and some ninety large and dangerous rocks jutting out of the water depending on the tide. All of the islands are small, at less than three miles (4.8 km) long. The partly submerged rocks occur around the islands and are great dangers to ships.

The *Torrey Canyon* did not have a fixed course so it could change directions as the captain saw fit. A supertanker of this size should have gone east of the Scilly Isles to avoid having to navigate a narrow twenty-mile-wide (32-km) passage between a spit of land and the Scillys. The captain risked the dangerous course because they needed to be in Milford Haven by the next day or it would take six days for the tide to be high enough for the tanker to enter the shipping channel again. The *Torrey Canyon* was cruising fast, near twenty miles per hour (32 km/h). As the tanker entered the area, it had to maneuver to avoid a group of fishing boats and entered the dangerous passageway. This was the first problem.

This navigation course had little margin for error. Unfortunately, a junior officer on his first trip made a navigational error by plotting the ship's location far from the Seven Stones shipping hazard even though it was only three miles (4.8 km) away. When the captain realized the mistake, he immediately ordered an emergency course change and reduction in speed. The problem was that unbeknownst to the captain and crew, the tanker was set to automatic steering so it did not respond. As a result, on the morning of March 18, 1967, at 8:45 a.m. and in full daylight in clear weather and a calm sea, the giant supertanker ran aground, hitting Pollard's Rock, a reef on the Seven Stones hazard between Land's End and the Isles of Scilly.

The collision tore a seventeen-foot-wide (5.2-m) gash across six oil storage compartments in the *Torrey Canyon*. This tear breached the entire tanker hold releasing the entire cargo of between 857,600 and 872,300 barrels, nearly 119,000 tons of crude oil, into ecologically sensitive coastal waters over the next twelve days.

This was the first accident and oil spill involving a supertanker ever, so no one knew how to deal with it.[7] The crew was evacuated and a salvage company was employed to fix the problem. They attempted to refloat the tanker by removing it from the hazard without success. They then tried to transfer the oil remaining in the supertanker to another tanker. However, at noon on March 19, one part of the tanker exploded, injuring five men and blasting two others into the sea. One of these men died and the other was rescued.

The next proposed solution was to have the British Royal Navy bomb the *Torrey Canyon* to ignite the spreading oil to contain and remove it. Over the following two days, about 62,000 pounds of explosives were dropped on the *Torrey Canyon* and surrounding waters from fighter jets, along with 5,200 gallons of gasoline, 11 high-powered rockets, and napalm. Of the 41 1,000-pound bombs dropped by fighter jets in "Operation Oil Buster," only 23 actually hit the tanker. Many of the bombs that missed the target caused significant ecological damage in addition to the damage from the spilled oil. The fighter planes were eventually successful in blowing the tanker off the rocks and sinking it. The bombing operation also ignited some of the floating oil and burned it.

Meanwhile, the oil slick from the *Torrey Canyon* wreck continued to spread.[8] With time it would spread out to cover between 270 and 386 square miles (700–1,000 km^2). It caused tremendous coastal pollution around Cornwall, UK, the Channel Islands, and Brittany, France, soiling some 120 miles (192 km) of the English coast and 50 miles (81 km) of the French coast.[9] In an attempt to stem this pollution, the British government dumped 11,230 tons of a crude oil dispersant in the ocean and on the shore. The application was so haphazard that, in some cases, barrels of the dispersant were just rolled off cliffs and allowed to drain. More than 1,400 British soldiers covered beaches with a mix of straw and a hay-like weed to recover oil. Nineteen days after the accident, oil reached the Guernsey's beaches. This oil was cleaned in an emergency operation that removed 3,307 tons of oil and sand that was dumped in a granite quarry on the island.

By late May 1967, floating oil slicks still persisted offshore of western Brittany. Winds continuously drove more oil toward the coast. In response, oil booms were constructed to protect harbors at Brest, Morgat, Douarnenez, and Audierne and nearby beaches. More than 3,000 French troops and volunteers used chalk mixed with stearic acid to recover oil on their beaches. They treated some 4,200 tons of oil waste but it took many weeks.

The spilled crude oil caused a terrible ecological disaster in the region that took several decades to recover.[10] It is estimated that between 15,000 and 30,000 seabirds perished in the toxic oil. It is estimated that there were more than 450 breeding pairs of razorbills before the spill but only 50 after it. Similarly, there were about 270 pairs of guillemots before and 50 after. About 85 percent of the puffin's population on the French coast also perished. One problem was that the well-intentioned volunteers washed the oil-coated birds with detergent to remove the oil. However, the detergent also removed the birds' natural oils that allowed them to repel water and retain body heat. Many of the rescued birds actually died of exposure. In addition, when the released birds preened, they ingested remaining detergent, which poisoned them.

Another major problem was that although the detergent dispersants were effective at breaking up floating oil, they were extremely toxic to marine animals. Later scientific studies showed that the dispersants were even more ecologically devastating than the spilled oil. Coastal rocky areas and beaches impacted by the spill recovered ecologically in about five to eight years. In contrast, the areas treated with dispersant took at least nine to fifteen years to recover ecologically, as much as five times as long as those that recovered naturally. Some areas where dispersants were heavily used or repeatedly applied had not fully recovered fifty years after the disaster.

Despite the massive cleanup efforts, a significant amount of spilled oil remained in the English Channel.[11] Some of the spilled oil was recovered and used for energy generation in the 1980s. However, it still periodically appeared on beaches and rocks for decades. In 2008, after a large soiling of the coast, oil-consuming microorganisms were released into the waters to consume the remaining oil. However, in 2009, an even larger quantity of oil appeared on the ocean surface and more extreme measures were employed. Government workers and volunteers removed about 42,268 gallons (160,000 l) of contaminated water using buckets. Even today, there remains an unknown amount of *Torrey Canyon* oil below the ocean surface. It remains as Britain's largest oil spill.

Atlantic Empress–Aegean Captain Collision

The *Torrey Canyon* was only the first in a series of phenomenal and disastrous ocean and coastal oil spills from supertanker accidents and many of these dwarfed it in terms of volume.[12] Because of *Torrey Canyon*, the first regulations on marine oil pollution prevention were initiated which evolved into the International Oil Pollution Prevention (IOPP) Certificate, which is required of oil tankers. For a single supertanker accident, the *Amoco Cadiz* incident was likely the largest, releasing 250,225 tons (227,000 mT) of crude oil to the waters around Brittany, France, on March 16, 1978. That caused a public health crisis in nearby towns from the soot generated by attempting to burn away the oil slick as well as the severe local ecological impacts. However, the biggest release of crude oil by supertankers was caused by a collision between two supertankers, the *Atlantic Empress* and the *Aegean Captain*, near Tobago in the Caribbean Sea.

The weather was terrible on July 19, 1979, off the coast of Tobago, in the southern Caribbean Sea.[13] Torrential rain and heavy fog shrouded the area, making travel treacherous that evening, even for ships. Unfortunately and fortuitously, there were two oil supertankers or VLCCs (Very Large Crude Carriers), transporting full loads of crude oil in the vicinity, each with different destinations. Both of these enormous vessels were less than twelve years old and equipped with modern navigational equipment, including radar and radio-directional indicators. Both also had well-trained crews and good safety records. Therefore, their passing should have been routine with no cause for alarm.

The *Atlantic Empress* was an enormous 1,139-foot (347 m) long by 170 feet (52 m) wide supertanker with deadweight tonnage of a massive 292,666 tons, making among the largest VLCCs at the time.[14] It was headed to the Mobil Oil terminal in Beaumont, Texas, with a full load of 1.9 million barrels (276,000 tons) of crude oil from Saudi Arabia. The tanker was built in 1974, so a mere five years old. Built in 1968, the *Aegean Captain* was older and a little bit smaller. The length of the supertanker was 1,076 feet (328 m) by 155 feet (47 m) wide and deadweight tonnage of 210,257 tons. It was heading in the opposite direction of the *Atlantic Empress*, bound for Singapore and carrying 1.4 million barrels (200,000 tons) of crude oil from Curaçao and Bonaire. The *Aegean Captain* was Greek owned but it was registered in Liberia.

At about 7 p.m., the two tankers were about eighteen miles (29 km) from the island of Tobago and visibility was near zero because they were in a tropical rainstorm.[15] Neither tanker was aware of the other or that they were on a heading that would lead them to collide. At about 7:15 p.m., the second officer of the *Aegean Captain* finally caught a glimpse of the *Atlantic*

Express when it was less than 600 yards (554 m) away. He sounded an alarm and ordered the huge tanker to immediately turn sharply away but it was too late to avoid the collision. The bow of the *Aegean Captain* struck a glancing impact into the side of the *Atlantic Empress*. There was a large explosion at the point of impact killing one crew member on the *Aegean Captain* and then both tankers caught fire.

Both captains ordered their crews to abandon ship. The evacuation on the *Aegean Captain* was carried out in an orderly manner as the fire was constrained to the bow of the ship. In contrast, panic ensued on the *Atlantic Empress*, which was in flames. Many of the crew members jumped into the ocean despite it being covered with burning oil and they perished in the flames. As a result, twenty-six crew members from the *Atlantic Empress* were killed in the disaster and five were hospitalized with severe burns.

The Trinidad and Tobago Coast Guard brought the fire aboard the *Aegean Captain* under control because all damage and flames were restricted to the starboard bow.[16] The disabled tanker was towed toward Trinidad by the tugboat *Oceanic*. Ten officers attempted to repair damage to the tanker during this time. However, the damage was more than anticipated and the *Aegean Captain* was towed to Curaçao. The tanker released an estimated 100,000 barrels of crude oil from the explosion and leakage as it was being towed. To prevent ecological damage, the tug sprayed dispersants on the spilled oil along the way and neutralized almost all of it. The oil remaining in the *Aegean Captain* was transferred to other tankers in Curacao.

In contrast, the *Atlantic Empress* remained ablaze and adrift. The starboard side of the main deck was engulfed in flames and the tanker was listing at about ten degrees. A two-mile-wide (3.2 km) oil slick extended about ten miles (16 km) from the damaged ship. A salvage team attempted to control the fire by spraying it with seawater from two tugboats while they towed the tanker out to sea. By July 24, the *Atlantic Empress* was 135 miles (217 km) from Tobago and still ablaze. That night, the tanker exploded again and began leaking oil at a much faster rate. The oil slick now extended fifty miles (80 km) from the ship but fortunately, it was still far from the shore.

The firefighting finally seemed to be working from July 25 to July 28 and the flow of leaking crude oil slowed. The team boarded the tanker and secured the water intakes and fire doors. The oil tanker *Tasso* was sent to unload the remaining oil from the damaged tanker. However, on July 29, the firefighters were forced to deal with the remaining blazes by shooting foam from guns and setting up forty-one hoses to spray seawater onto the decks. They were making good progress and planned to return the next day. However, a few minutes after they departed, there was yet another explosion on the *Atlantic Empress*. The flames were 300 feet (91 m) high and 35 to

70 feet (11–21 m) of the deck had been blown open. Flaming oil extended at least 200 yards (185 m) from the tanker. Most of the firefighting equipment had been destroyed by the explosion and the fire began spreading to other parts of the tanker.[17]

By August 2, the *Atlantic Empress* was leaking crude oil at an even faster rate of 7,000 to 15,000 gallons per hour (26–57 m³/h). The tanker was engulfed in flames as the leaking oil pooled around it and was ignited. As a result, the tanker began to sink with the red hot bow sticking up out of the water. Nearly two weeks after the collision, the big tanker sank in several hundred feet of the Caribbean Sea. As it went down, the flames still shot 500 feet (152 m) into the air and the smoke wafted 6,000 feet (1,829 m) into the air. The fire burned most of the remaining oil but the slick from the release was about 30 miles (48 km) by 60 miles (96 km). However, the slick did not reach any nearby islands and aircraft sprayed dispersants on it until, by August 9, it was no longer visible.

The combination of spills from the *Aegean Captain* and *Sea Empress* was a whopping 316,363 tons (287,000 mT) of crude oil, the record spill into the ocean from supertankers. This was only one year after the huge spill from the *Amoco Cadiz* accident of 250,225 tons (227,000 mT) which held the record at the time. However, because the main destruction of the *Sea Empress* happened so far from land, and much of the spilled oil was treated immediately, the outward ecological impacts appeared relatively minor. Some soot from the burning oil contaminated a few drinking water supplies in Tobago. The oil slick, however, did not reach land. The most toxic chemicals immediately evaporated from the thin layer of oil in the hot sun. Much of the oil was dispersed in the seawater after being helped by the chemical dispersants from tugboats and aircraft. The damage to marine plankton and other sea life is unknown.

Exxon Valdez Disaster

Even though it was not close in volume to the record largest spill, the *Exxon Valdez* supertanker accident may have been the most impactful in the United States and perhaps the world in terms of being the impetus for regulations. It was the beginning of the end for the ever-increasing ocean oil pollution. The tanker was not as large as those in the previous examples. It was 987 feet (301 m) long, 166 feet (51 m) wide, and 88 feet (26 m) deep and had a 240,644 deadweight tonnage (218,309 mT).[18] The tanker had the capacity to carry 1.48 million barrels (238.5 million L) of crude oil. The tanker was as modern as available at the time with a strong uniweld hull, ten cargo tanks, and four ballast tanks providing maximum stability and cargo management.

The *Exxon Valdez* tanker began its shipping life on December 6, 1986.[19] Its sole purpose was to transport Alaskan crude oil from the Port of Valdez to refineries in California. Valdez, Alaska, was home to the distribution end of the 800-mile (1,288-km) Trans-Alaskan Pipeline, which transported crude oil from Prudhoe Bay on the Alaskan north slope to market. Valdez is a relatively ice-free harbor so shipping oil takes place all year. ExxonMobil owns a 20 percent interest in the pipeline. The pipeline was planned for a long time but environmental concerns slowed its approval. However, the OPEC oil embargo of 1973 panicked the country enough to remove any obstacles and the pipeline was built between March 1975 and May 1977 and it was operating by July 28, 1977. Loaded tankers were leaving Valdez Harbor by August 1977 and the number reached about seventy tankers per day during peak activity.

The disaster began on March 22, 1989, when the empty *Exxon Valdez* tanker arrived at Port of Valdez for loading at 11:30 p.m.[20] The captain, Joseph Hazelwood, had almost twenty years of experience with a solid record and the *Exxon Valdez* was the newest and most modern tanker in the Exxon shipping fleet. The tanker was loaded with 1.26 million barrels (53 million gallons) of crude oil from the pipeline.[21] The *Exxon Valdez* supertanker departed the port at 9:12 p.m. and ferried south under the control of the harbor pilot. It cruised through the tight, one-mile-wide (1.6-km) entrance to the harbor and out into the open waters of Prince William Sound.

Prince William Sound opens into the Gulf of Alaska which is open water and about 100 miles (160 km) wide. The sound is rimmed by about 3,000 miles (4,828 km) of shoreline and it contains small islands, submerged reefs, and fjords that form an important habitat for numerous seabirds, marine life, and commercial fisheries. Much of the area around Prince William Sound was designated as a national forest in 1907. The Chugach National Forest is a large part of this land and it is the second-largest national forest in the United States. In 1974, the US Congress passed the Alaska National Interest Lands Conservation Act largely to preserve this and similar areas.

Once the *Exxon Valdez* cleared the Valdez Narrows, the harbor pilot departed and turned control back over to the Exxon crew at 11:24 p.m. but not before Captain Hazelwood abruptly retired to his cabin.[22] The sequence of events becomes somewhat confused after this depending on the account. The tanker apparently entered the clearly marked southbound shipping lanes which lie along the west side of Prince William Sound. The weather was clear and calm with four miles (6.4 km) of visibility so there were no weather issues. The tanker speed was fast at about twenty knots (37 km/h) and a request was made to change its course eastward into the northbound lanes to avoid some floating ice at 11:30 p.m. The port traffic control granted permission and the tanker steered eastward and into the northbound lanes. The plan was to pass through a one-mile (1.6 km) gap in the icebergs and between a group of shallowly submerged rocks. The supertanker was planned to then turn back into the southbound lanes once through the ice hazard. This complex maneuver required expert navigation for such a huge ship but the bridge was now under the control of a single, young third-mate, who didn't have the clearance to pilot the tanker in addition to it being against company policy. Another officer was scheduled to take control around this time, adding to the confusion.

The junior officer was left with instructions to return to the southbound lane when he spotted the navigational beacon on Busby Island in the sound. His attention, however, was both on watching for the ice floes on radar and the signal light indicating that he should turn. As a result, he apparently missed seeing the Busby Island light. He did not realize the mistake until a crew member reported the rapidly approaching lighted buoys that marked the Bligh Reef and there was some confusion as to where the tanker was located at the time. In a panic, he immediately ordered a sharp evasive maneuver but it was too late. The *Exxon Valdez* supertanker drove 600 feet (182.9 m) into the massive rocks of Bligh Reef at nearly full speed at 12:04 a.m. on March 24, 1989. Several of the crew reported feeling six bumps. Captain Hazelwood was being called back to the bridge just as the collision was occurring.

The momentum of the tanker drove it to the point that it was perched on a pinnacle of one of the rocks.[23] The scraping over the rocks punctured eight of the eleven cargo holds. Back on the bridge, Captain Hazelwood made numerous attempts to dislodge the huge tanker from the rocks by rocking it but all he did was make the holes larger and increase the damage. He gave up this attempt and radioed the Coast Guard to report the accident. Between the collision and fruitless efforts to free the *Exxon Valdez*, the tanker bottom was ripped wide open and oil poured into the sea. Within thirty minutes, at least 110,000 barrels of crude oil (17.4 million L) spilled into Prince William Sound. After eight hours, the spill was more than 215,000 barrels (34.1 million L) and over the next two days, it reached 257,000 barrels (41.6 million L).

The accident was the first actual implementation of the response plan for an oil spill in the Port of Valdez or Prince William Sound. The first of many problems in the rescue was that emergency response ship was in long-term dry-dock because of damage it sustained several months earlier. Second, the emergency oil spill containment booms and skimming equipment were packed away in a warehouse and there was only one equipment operator available to operate the forklift and crane to load them onto the emergency response ship. Nonetheless, in spite of a crack in its bow, the response ship was loaded and dispatched to the spill. As a result, it took more than fourteen hours for the response ship to reach the accident. By then the slick had covered nearly twenty square miles (51.8 km²) and was expanding rapidly.

Third, pumping the oil that remained in the *Exxon Valdez* into other tankers to remove and save it was delayed for fear that shifting weight might capsize and sink the damaged ship. Also, the pumps necessary to transfer the oil from the tanker could not be located. The fresh crude oil was so close to the shoreline that officials were reluctant to spray dispersants because there could be long-term damage to the ecology. This was somewhat acceptable, at first, because the calm weather was preventing the spill from spreading. However, the weather changed and strong winds and rough seas developed, spreading the oil all along the shore.[24] It covered at least 1,300 miles (2,092 km) of coastline by the end of the incident.

The *Exxon Valdez* accident and spill was well publicized by the media to an outraged American public. The newspapers and evening news were awash with photos of oil-coated dying seabirds and otters. Exxon, however, took full responsibility for the disaster and attempted to remediate the damage. They hired more than 11,000 personnel and dispatched 58 aircraft and at least 1,400 vessels to perform the extensive cleanup during the summer of 1989. There were also many volunteers both locally and from other areas. The cost of the cleanup was estimated at $2 billion in 1989. A number of methods were used to remove the oil such as limited burns and dispersants in addition to mops, buckets, pressure washers, and steam cleaners. The pressure washing damaged the sensitive local ecology. Environmental officials left some segments of shoreline untreated to study the impact of the cleanup measures. They determined that washing with high-pressure, hot water hoses was effective at removing oil, but it did extensive ecological damage, killing all the plants and animals it encountered.

The massive cleanup effort had mixed results. At least 660,000 gallons (2.5 million L), or 6 percent of the spill, were recovered by skimming, some 1.9 million gallons (7.2 million L) or 18 percent evaporated or photodegraded in the sunlight, 3.1 million gallons (11.7 million L) or 28 percent were emulsified and dispersed into the seawater, and a whopping 5.3 million gallons (20.1 million L) or 48 percent of the spill washed up on the shoreline of Prince William Sound and decimated the local ecology.[25] It is estimated that more than 250,000 seabirds were oiled and killed. The same fate befell 2,800 sea otters, 12 river otters, 300 harbor seals, 250 bald eagles, and as many as 22 killer whales. The oil and dispersants also destroyed billions of salmon and herring eggs. Fishing for both salmon and herring was banned the next year, and the populations have never fully recovered.

The oil spill reduced the populations of cormorants, goldeneyes, mergansers, murres, pigeon guillemots, and other marine birds for more than nine years after the disaster.[26] The oil killed about 40 percent of the sea otters in the sound and the population did not recover to pre-spill levels until 2014, some twenty-five years later. Mussel beds and several other tidal shoreline habitats took thirty years to recover. There are only a few traces of the spill still visible to the casual observer today but there is still a layer of oil present just a few inches below the surface in several locations. The ecology in several areas has also still not recovered nearly thirty-five years after the accident.

A thorough investigation concluded that the direct cause of the accident was the inexperienced officer's failure to make the course correction as ordered. The situation was exacerbated by his fatigue as he had supervised the oil onloading that day and only had a few hours of sleep. The other major issue identified was the criminal lack of command on the bridge. As the investigation revealed, Captain Hazelwood had been out drinking prior to leaving port and did not leave the bridge in proper hands or with proper navigation plans and assurances. This was not the first time that Captain Hazelwood had issues of substance abuse at unwise times. He should have been relieved of command for these previous incidents and Exxon had knowledge of his problem but opted to take no action. The final contributing issue identified in the investigation was that the Coast Guard and the Port of Valdez traffic control radar installations were old and not up to the task as well as being inadequately staffed.

In response to these public findings, the culture at the huge Exxon Corporation was completely shaken up to zero tolerance of any alcohol or perception-altering substance of any kind.[27] The Coast Guard replaced its radar equipment and system of operation in Prince William Sound. They imposed new shipping regulations that strictly forbid ships from changing lanes, except in the case of an emergency and set a strictly enforced ten-knot speed limit. Tanker captains industry-wide now must pass a mandatory sobriety test one hour before departing port. Captain Hazelwood was taken up on criminal charges. He was found guilty and convicted for negligent release of oil. As a consequence, he had his Masters License suspended, he was fined $50,000 and he was sentenced to 1,000 hours of community service. He served this sentence over the next five years working in a soup kitchen and picking up trash along several Alaskan highways.

The findings and actions, however, did not end the woes for Exxon. In 1994, an Alaskan court convicted Exxon in the incident and required them to remit $5 billion in punitive damages. This action was met with several appeals by Exxon who claimed they had already paid enormous sums. Finally, the case went to the US Supreme Court who reduced the charge to $507.5 million which Exxon paid. It has been calculated that Exxon spent at least $3.8 billion on the cleanup and compensated more than 11,000 fishermen and other people and businesses that were impacted by the disaster. In addition, at least 600,000 acres (242,811 ha) of land around Prince William Sound are now protected through utilizing settlement and matching funds from the numerous restoration, research, and monitoring programs that surrounded the disaster and cleanup.

OIL POLLUTION AND CONTROL ACT

There were significant changes to oil transport using supertankers that were direct results of the *Exxon Valdez* disaster but nothing compared to the US government response, which was the passage and enactment of the Oil Pollution Control Act of 1990.[28] This was not the first legislation governing oil spills. The US Limitation of Liability Act of 1851 is actually broad enough to include oil spills even though there would not be any oil spills for many years after its passage. This act held shipowners liable for all costs related to a ship accident up to the value of the ship after the incident. This may have been acceptable for many previous ship accidents but the legislators never imagined the damage that could be inflicted by a crude oil spill from a supertanker. Although there were subsequent acts that regulated purposeful and accidental ocean dumping, they did not cover oil spills. In fact, the US Congress enacted the Ocean Dumping Ban Act of 1988 just months before the *Exxon Valdez* disaster showing that there was already national concern over ocean health. The Ocean Dumping Ban Act makes it unlawful to dump, or transport to dump, sewage or industrial waste into US ocean waters after December 31, 1991.

The Oil Pollution Control Act of 1990 was passed by the US Congress and signed into law by President Bush. The law intends to prevent oil spills from ships and facilities by requiring removal of the spilled oil and covering the full cost of the cleanup and damage caused by the responsible parties. It requires certain operating procedures, defines responsible parties, outlines procedures to measure damage and assign liability, and it established a fund to cover costs of cleanup and damage. The law requires any tanker operating in US waters to be built with a safer reinforced double hull. All new tankers are built with these and other safety improvements as a result of this law. The law also prohibits any ship that spilled more than 1 million gallons (3.8 million l) of oil in any marine area worldwide after March 22, 1989, from ever entering Prince William Sound.

The result of this law and the accompanying safety regulations was a marked decline in accidental spills of oil in the marine environment by tankers.[29] Recent data show that accidental tanker spills of crude oil are down 90 percent and the volume is down 95 percent on a worldwide basis compared with the 1970s. They also show that 99.9 percent of crude oil was delivered safely by tankers in 2022. This is not to say that there have not been any tanker accidents and spills. There was a tanker collision in the South

China Sea on January 6, 2018, in which 152,119 tons (138,000 mT) of crude oil was spilled into the ocean. This is more than either the *Torrey Canyon* or the *Exxon Valdez* showing that the threat is not completely gone. In addition, oil can be spilled into the ocean by other means, such as oil well blowouts. The BP Deepwater Horizon blowout and oil spill released as much as a phenomenal 691,150 tons (627,000 mT) of crude oil into the Gulf of Mexico between April 20 and July 15, 2010. This is far in excess of any supertanker spill and possibly the largest spill ever. Fortunately, these kinds of disasters are few and very far between. Most recent crude oil spills are from pipelines on land which can also be large and destructive as well. This problem remains in desperate need of a solution.

THE TAKEAWAY

Oil supertanker accidents and releases, especially in coastal regions, progressively increased from the late 1960s through the 1980s. They may not have caused an inordinate number of human deaths or worldwide damage to specific species of plants and animals but each one devastated the local marine ecology of the impacted areas. As the number of accidents and amount of crude oil released to sensitive ecological settings severely damaged area after area, they began to have cumulative effects. If action had not been taken to reduce these accidents and volumes released, more widespread permanent

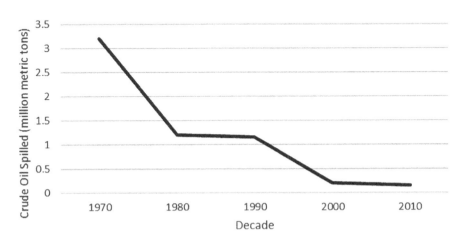

Figure 6.1.   Graph showing total tonnage of crude oil spilled in tanker accidents per decade from the 1970s through the 2010s.
Source: Data from International Tanker Owners Pollution Federation (ITOPF).

damage could have resulted and several species would have been threatened or driven to extinction. However, public outrage and resulting pressure forced federal action to reduce or remove the threat. The 90 percent reduction in supertanker spills and 95 percent reduction in total volume of spills means that the threat has been nearly removed. This is a huge victory over a serious environmental threat.

Chapter 7

Burning Rivers and Urban Surface Water Pollution

Surface water is even more of a dumping ground for society than the air. In standing water bodies like lakes, most waste just sinks and disappears from view. Rivers are even better because if waste is dumped into the flowing water, it is swept away and out of view. Normally, this system of flushing away waste works well. The flowing water disaggregates the waste and dilutes it as well as removing it. Problems only develop when a permanent human settlement is constructed and especially as it grows larger. The basis of many problems is the conflict of using the river as both a source of water and food as well as for waste disposal. Over time, people learned to use the river water as it enters the settlement and dump waste into it as it exits. As settlements matured into cities, there have been numerous cases of public sewers emptying raw sewage into rivers on the downstream sides of the city.[1] The next city downstream of this sewage disposal therefore had their river water quality impacted. But they further polluted the river water with their own waste that they discharged downstream of the city. It has been speculated by some that by the time Mississippi River water reaches New Orleans, it has been used six times. With this practice and the huge amount of waste dumped and washed into the river within the city, urban rivers became putrid.

SEWER SYSTEMS AND RESERVOIRS
FOR CLEAN WATER

This problem of raw sewage contaminating river water and causing public health problems is as old as civilization. The primary issue is bacteria and other pathogens infecting large numbers of residents and causing outbreaks of diseases like cholera, dysentery, hepatitis A, typhoid, and polio, in addition to regular, widespread diarrhea.[2] One solution to this problem is to develop sanitation systems to remove the waste. This practice began early in settlements in the Indus Valley in now Pakistan and northwest India between about 3300 and 1300 BCE.[3] They developed among the most advanced sanitation systems in the ancient world. Another solution is to develop a separate source of fresh water with canals, water tunnels, cisterns, and even aqueducts. The cleanliness is really a secondary reason to supply enough water for the settlement population. For this reason, these additional water supply systems were first well developed in drier areas like ancient Persia and Syria, mostly in the last millennium BCE.

Ancient Rome

The premier public water supply systems and sewer systems of the ancient world were in Rome and they were not equaled again until the nineteenth century in the industrialized world.[4] However, it did not start out that way. For the first 400–450 years, Romans obtained water from the Tiber River which flowed through the center of the city or from springs, wells, and tributaries. However, it became clear that the Tiber River was highly contaminated with raw sewage by the end of this time. There are reports of carcasses of dead animals, fecal matter, and all kinds of solid waste littering the river to the point that some emperors undertook large-scale projects to clean it. At times, it was difficult to find drinkable water.[5] Some estimate that the mortality rate of children under ten years was as much as 50 percent, largely as the result of waterborne disease. Studies of Lake Murten in Switzerland show that waste from a nearby Roman town caused serious eutrophication of lake water.[6] Waste in Rome was far more extensive.

To overcome the lack of fresh water, the Romans built a spectacular aqueduct system that transported vast quantities of clean water to the city.[7] The population of Rome was approximately one million people at its peak which required a lot of water. It was not only used for drinking but also for the extensive baths in the city, industry, and fountains among other uses. However, this did not relieve the Tiber River of the raw sewage and other waste or the reported waste that littered the streets and even homes and buildings.[8] It took a few centuries, but Rome finally constructed the Cloaca Maxima, the first effective and expansive sewer system in the city. This further relieved the pollution and resulting disease. These systems were the envy of the ancient world.

New York City

Just because Rome's waste issues were 2,000 years ago doesn't mean humans don't have the same issues and solutions in modern times. New York City is a great example of not relying on local water but instead piping it in from large reservoirs quite a distance away. The settlement started out using local wells and ponds for water but it could not use local rivers because they are all brackish.[9] They were used for waste disposal. This salty water even contaminated wells that were sited too close to the rivers. The lack of a plentiful source of fresh water quickly became impossible so Manhattan began importing fresh water from Brooklyn by the early 1700s. But even this solution was short-lived as the population continued to grow. The shortage of fresh water meant that fires could not be extinguished and they ravaged the city on a regular basis as a result. In 1776, a single fire destroyed about one quarter of the buildings in New York. Local water was contaminated by waste but was still used, causing numerous health crises. In 1832, there was a cholera epidemic that killed 3,500 people, and it was blamed on contaminated water.

New York had to come up with a better system. In 1799, they contracted the Manhattan Company headed by Aaron Burr to bring in water from outside of the city. But instead, the company just drilled more wells and built a new reservoir downtown and then focused on banking, eventually becoming Chase Manhattan. The result was that the water supply situation got even worse. In the late 1820s, the city finally bought land north of its populated area and in 1837, built the Croton Reservoir in Westchester County and piped in clean water forty-one miles (66 km) by aqueduct.[10] Fresh water began flowing into the city by 1842 and resolved the water quality and shortage issues, at least temporarily.

By the 1880s, the touted system was already inadequate. The city greatly enlarged the Croton reservoir system and its watershed by purchasing more land and building bigger dams. In 1885, construction began on a water tunnel that could transmit nearly three times as much water as the old Croton reservoir aqueduct. However, this greatly increased water supply was also quickly exhausted by the widespread installation of household plumbing, construction of an extensive sewer system, and an even more accelerated population growth.[11] In 1898, when the boroughs were merged into the single modern city of New York, the water supply immediately became inadequate once again. Within a few years, New York City began purchasing land well to the north in the sparsely inhabited Catskill Mountains to build the huge Catskill reservoir system which was operational by 1927. The Catskill aqueduct and water tunnel transport water ninety-two miles (148 km) south to the city. But this still wasn't enough fresh water. In 1965, New York City opened a third Delaware reservoir system eighty-five miles (137 km) north of the city. It transports water to the city through the longest tunnel in the world. This is not even the end of the work as New York City is still constructing water tunnel number three and will be until 2032.

The three New York City watersheds and reservoir systems cover an area of nearly 2,000 square miles (5,180 km²). This is about the size of the state of Delaware.[12] The reservoirs have a combined capacity of about 550 billion gallons (2,082 billion L) and supply 97 percent of the water to the city. The water is delivered to the city through aqueducts and pipes that are more than 6,200 miles (9,978 km) long. This system is the best illustration of what lengths communities will go to for clean water.

These improved water and sewer systems also improved the water quality and decreased the mortality rate in many American cities. However, another major improvement to water quality was the addition of filtering to remove unwanted compounds and sediment and chlorine to remove pathogens. These practices were added in American cities between 1900 and 1940 and greatly reduced the incidence of waterborne diseases like typhoid and cholera. This accounted for about half of the approximate 30 percent decline in urban mortality rate over the period across the United States.

PROFOUNDLY POLLUTED SITES

Hackensack Meadowlands

The surface waters right across the Hudson River from New York City, however, probably suffered the worst pollution in the United States, if not the world at roughly the same time as New York City was addressing their water problems.[13] These waters include the Passaic River which serves Newark, New Jersey, among other cities, the tributary Hackensack River and all of the surrounding wetlands and tributaries of the Hackensack Meadowlands or simply the Meadowlands. In addition to being suspected as the final resting place of union boss Jimmy Hoffa, this current estuary and the surrounding wetlands formerly comprised a large glacial lake. Being so close to New York City, this area has always been an important location. As a result, the local residents have attempted to convert this swamp to usable land by infilling it and building canals and dikes. This altering of the landscape caused the incoming rivers to silt up, reducing the freshwater input thereby allowing in more saltwater and causing the chemistry of the Meadowlands to become more saline.

Serious pollution began in the Passaic River in the nineteenth century, when the raw sewage and industrial wastes from Newark and surrounding areas were being dumped into it.[14] It got so bad that by the end of the century, pollution began backing up into the Hackensack River and southern Meadowlands. The Passaic River got so polluted that, in the early 1900s, Newark and several other cities built a long sewer line that pumped sewage directly into the saltwater of Newark Bay off of the Atlantic Ocean. This somewhat reduced the pollution in the Passaic River but Newark Bay became severely polluted. The tides in Newark Bay then drove this polluted water up the Hackensack River more than twenty-one miles (33.8 km) into the Meadowlands. The Passaic River was dredged at about the same time to deepen and widen it, which allowed even larger amounts of increasingly polluted tidal waters from the Newark Bay and New York Harbor into the Hackensack River and Meadowlands.

At this time, a movement was underway to reclaim wetlands from the Meadowlands and most of eastern New Jersey.[15] It was being done using fill made from mixing urban garbage with solid materials like rocks, gravel, ash, bricks, and concrete. A lot of the garbage was shipped in from New York City which was struggling with the vast amount of garbage it was generating. One of the fill projects was in 1914 to construct Port Newark. A shipping channel was dredged from Newark Bay into the Hackensack River. The dredged sediment was mixed with garbage and ash and dumped in the Meadowlands to make new land a few feet above sea level. Docks and warehouses were then built on it.

Soon the Meadowlands wetlands were having more waste dumped in them than anywhere else in the United States.[16] During World War II, military waste was dumped in the Meadowlands. After the war, it began being used for a truly massive disposal site for municipal waste. The New Jersey Turnpike was built through the Meadowlands in 1952, making transport of waste into the area much easier. Tons of garbage was deposited in open dumps and along roadsides illegally and at will. The expanding road system at the time provided easy access to trucks out of New York City and New Jersey communities and into the Meadowlands. Lawlessness prevailed in this mess and mobsters were free to control the areas they wished and to charge unregulated "tipping fees" to the trucks that entered the region. This uncontrolled open dumping peaked during the 1950s and 1960s and turned the Meadowlands into the world's largest landfill at the time. The bigger dumps grew into large mounds which were regularly burned to make more space for garbage and to control wind-blown debris. The methane gas being produced by the rotting garbage along with other flammable debris regularly caught fire and burned uncontrollably for years.

In 1957, New York City stopped providing garbage removal for manufacturing companies located in the city. The companies were forced to hire private garbage collectors, many of whom were associated with organized crime.[17] These enterprises eliminated their disposal costs by dumping this waste at any unmonitored location. Other crime enterprises even set up unregulated toll booths along the main roads to collect tipping fees for dumping by both criminal and legitimate companies during this time. The highway access, low population density, and minimal law enforcement in the northern Meadowlands made this possible. The local waste disposal industry saw explosive growth at this time. Multiple garbage haulers made nearly continuous runs to ever-increasing numbers of illegal dumps which attracted the attention of residents, newspapers, and finally law-enforcement officials. They made feeble attempts to control the influx of waste at public urging with marginal success.

During the 1960s, at the peak of dumping in the Meadowlands, it received 40 percent of the solid waste generated in New Jersey as well as a reported 10,000 tons (9,072 mT) of waste per day from New York City. There were at least 200 waste dumps that occupied more than 2,500 acres (1,012 ha). The burning of this waste polluted the air with foul-smelling smoke which mixed with the stench of decaying garbage and methane emitted from landfills.[18] The noxious odors sickened residents and even motorists on the New Jersey Turnpike. This scene alone gave New Jersey a bad reputation that persisted for decades. In the early 1970s, the smoke from burning waste got so thick that it caused a massive pileup traffic accident on the New Jersey Turnpike. Organic liquid waste or leachate from landfills, toxic chemicals, and raw sewage from surrounding municipalities polluted the Hackensack River to the point that eutrophication and resulting fish kills were commonplace.

Sewage and municipal waste were not the only things being dumped into the Meadowlands.[19] Industrial toxic waste was also being deposited by several companies creating profoundly polluted sites. The Ventron/Velsicol Company operated a mercury processing plant on a forty-acre site from 1929 to 1974. Mercury was removed from batteries, lab equipment, and other devices and much of it was dumped on the property. At least 160 tons (145 mT) of waste was buried in the mud which yielded mercury concentrations up to 195,000 parts per million (ppm) in soil. There are layers of pure liquid mercury in the soil. They also dumped waste into Berry's Creek, a tributary of the Hackensack River, and it spread several thousand feet downstream. An estimated 268 tons (243 mt) of mercury waste was discharged into the creek between 1943 and 1974. As a result, sediments in the creek have one to two parts per thousand of methyl mercury, the highest concentration in freshwater sediments in the world. Methyl mercury is toxic and contaminates all fish through bioaccumulation. Even the ambient air around the site has elevated mercury.

The Standard Chlorine company constructed a chemical manufacturing plant on a forty-two-acre (17 ha) site in the Meadowlands that was later operated by several companies from 1916 to 1993. The plant refined naphthalene, processed liquid naphthalene, and manufactured lead-acid batteries, drain-cleaner, and dichlorobenzene packing products. They dumped toxic waste containing dioxin, benzene, naphthalene, PCBs, and volatile organic compounds (VOCs) into the soil, groundwater, and two large waste lagoons. The lagoons drained this waste into the Hackensack River. There were also storage tanks and numerous drums full of hazardous substances such as dioxin and asbestos stored on the property.

The Diamond Head Oil Refinery included an oil reprocessing plant on a 20.2-acre (8.2 ha) property in Kearny, New Jersey. It was operated by PSC Resources, Inc., Ag-Met Oil Service, Inc., and Newtown Refining Corporation from 1946 to 1979. The property contained numerous storage tanks and several surface pits where oil waste was stored. This waste was periodically dumped into the wetlands which formed a large "oil lake." As a result, the property was highly polluted with light nonaqueous phase liquid (LNAPL), dioxin, chromium, PCBs, lead, aldrin, and thallium in both the soil and sediment.

Perhaps the worst polluted site was on the Passaic River. The Diamond Alkali Corporation severely polluted a 17.4-mile (28 km) stretch of the Passaic River, the lower Hackensack River, and into the Newark Bay. The plant began operations in the 1940s but it greatly expanded in the 1950s and 1960s, when they manufactured agricultural chemicals, including herbicides used in the Vietnam War defoliant "Agent Orange." Agent Orange contains dioxin, an extremely toxic chemical. The soil and groundwater at the site, in the river, and in the river sediments contain dioxin and several other hazardous substances including PCBs, heavy metals, polycyclic aromatic hydrocarbons (PAHs), and pesticides. As a result, the fish in both the Passaic and Hackensack Rivers have very high levels of bioaccumulated dioxin and cannot be consumed.

The state of New Jersey finally intervened in 1969 by attempting to halt this uncontrolled pollution in the Meadowlands.[20] At that time, at least 5,000 tons (4,535 mT) of garbage from 118 New Jersey municipalities was being dumped in fifty-one landfills, six days per week. The state established the Meadowlands Commission to control and oversee the sanitary disposal of solid waste. Sanitary landfill disposal of municipal waste was slowly reduced and continued in the Meadowlands until 2020. Now, no solid waste is being dumped into the Meadowlands and the job of remediating this profound pollution is underway.

Cuyahoga River Fires

Perhaps the most infamous case of extreme urban river pollution was on the Cuyahoga River.[21] The Cuyahoga is the major river in Cleveland, Ohio, and it flows into Lake Erie at its north end. It was a key feature in the development of the city. By the early 1930s, the river had been straightened, widened, and deepened by the US Army Corps of Engineers to facilitate large ships entering Cleveland from the lake. It was so heavily industrialized that by 1936, large parts were so polluted that it could not be used as cooling water by factories and even made transport of goods by barge difficult to impossible.[22] Flammable pollutants increased to the point that the river caught fire on numerous occasions. River fires began by the 1890s in a number of major American cities but fires on the Cuyahoga River were the worst. They were more frequent and more intense than virtually all of the others.[23] Petroleum products and flammable debris in the river caught fire and produced infernos in 1868, 1883, and 1887. In 1916, the sparks from a passing tugboat ignited oil that had leaked from a refinery on the river, setting off several massive explosions and a raging inferno that killed five workers. In response, Cleveland enacted a law that prohibited the release of oil or other flammable products into the river by refineries. In spite of this, fires occurred a few years later in 1922 and again in 1930. In 1936, a major river fire could not be extinguished for about a week.

In the same year, a major fire began near a tugboat facility that burned for six hours and produced flames more than five stories high which damaged a bridge, docks, piers, and barges. The fuel for the fire was primarily unused and waste oil that was dumped into the river as well as industrial and municipal trash and debris. Steel mills, oil refineries, and other industrial plants along the river dumped untreated wastewater, unrefined or unusable petroleum products, and a variety of organic chemicals into the polluted water on a daily basis. Even the city of Cleveland and surrounding municipalities released raw sewage into the river continuously for more than 100 years.[24] Oil and deicing chemicals from roads and fertilizer and pesticides from agricultural areas were swept into the river by runoff during rainstorms and spring thaw. As a result of fighting all of the river fires, the Cleveland City Fire Department became the national expert. The department owned advanced firefighting tugboats and their crews regularly extinguished Cuyahoga River fires. They were experts at quickly spotting oil slicks and applying chemical dispersants to them.

After a major river fire in 1952, residents of Cleveland became frustrated enough to pressure city officials to address Cuyahoga River pollution. Their response, however, was feeble and a decade later the river was still heavily polluted and clogged with waste. In 1962, they could only manage to keep the main channel open for barge and other river traffic, which further infuriated the residents. In response, in 1963, the city government created the Cuyahoga River Basin Water Quality Committee which immediately began a program to remove debris from the river and investigate several major point-source polluters. This committee began a water quality surveillance program to identify polluters.

Cleveland began to lose its industrial base and, as a result, its population by the late 1960s. By then, the residents and nation had become very concerned about environmental damage. Consequently, a $100 million environmental bond issue was approved by the voters in 1969, primarily for the cleanup and revitalization of the Cuyahoga River and its watershed. The bond issue included aggressive debris removal and antidumping enforcement programs as well as major improvements to sewage treatment plants along the river course. Another program that was begun in 1969 introduced methods to prevent new waste oil from entering the river and to remove the existing oil.

In spite of all these efforts, on June 22, 1969, the river once again caught fire.[25] The blaze was quickly brought under control by the fire tugboat crews with only minimal damage to the surrounding area. Nonetheless, it emerged as a seminal environmental event in the American environmental movement. First, it embarrassed the city of Cleveland and reinforced its reputation as an environmental wasteland that persisted for many years. However, more importantly, the 1969 fire attracted the attention of the national media.[26] The problem was that the fire was extinguished before the news film crews arrived so they got no recordings of it. Their solution was to use photos and news-reels of the much larger and more destructive 1952 river fire in newspapers and on television which shocked the already enraged public. As a result, the Cuyahoga River fire was included with the Santa Barbara oil well blowout and several other highly publicized environmental disasters as the impetus for federal action on environmental issues—whether it deserved it or not.

CLEAN WATER LEGISLATION

Clearly, even in modern industrialized countries, pollution of urban rivers quickly developed into a public health and even safety threat. This threat was quickly recognized and cities attempted to take action at least with regard to providing fresh water sources and sewers. In the United States, the Federal Water Pollution Control Act of 1948 was enacted as the first major legislation to address water pollution but it was largely ineffective. It took the public fervor of the 1960s to force enactment of truly effective legislation. The first step was the establishment of the US Environmental Protection Agency in 1970. However, its first charge was to enforce the 1970 Clean Air Act which had been enacted that year. It would take until 1972 to enact the Clean Water Act which would finally allow urban rivers to be cleaned up but it took some pressure and negotiation to see it through.

The Clean Water Act was passed by the US Congress and Senate and sent to President Nixon to be signed into law.[27] However, on October 17, 1972, President Nixon vetoed the Clean Water Act. He claimed to be in favor of it but thought it cost too much. When it was returned to the Senate, the sponsor of the bill, Senator Muskie of Maine, admonished his colleagues on the necessity of the bill for the health and well-being of the nation. His impassioned speeches convinced them to work together to make this important bill pass. By the next day, both the Congress and Senate voted to override the president's veto and the Clean Water Act became law.

This was a huge victory for the environmental movement. Politicians from both sides of the aisle came together to defy a sitting president. It reflected the strength of the resolve of the American people. It also reflected that the movement had been influential enough that the elected officials placed the health and safety of the public over corporate priorities and greed. It illustrates what is possible when citizens exercise their rights.

The Safe Drinking Water Act required the EPA to establish standards to restrict individual contaminants in drinking water.[28] They set standards for about ninety contaminants but they also studied and continue to study unregulated contaminants to determine if they should be regulated, as well. This effort requires monitoring, reporting, and other requirements of the quality of the public water supplies across all states with reporting to the EPA. The Clean Water Act also governs the pollution in surface waters including urban rivers. It includes pollution control and developing standards for surface water pollutants.

Before 1972, companies and municipalities could legally release wastewater into surface waters without even testing for pollutants much less removing them. The Clean Water Act allowed the EPA to establish the National Pollution Discharge Elimination System which issues permits for any discharge of wastewater into surface water bodies.[29] The EPA requires every facility to obtain a permit specific to the type of wastewater and to treat it before releasing it. There are monitoring and reporting requirements to ensure that the pollution standards are met. There are also additional or more stringent requirements specific to certain states and areas for protected water bodies. These regulations apply to point sources of pollution which can be uniquely identified and addressed but there are also regulations for nonpoint sources of pollution in some cases. Nonpoint sources cannot be uniquely identified because they are from diffuse sources like road chemicals, lawn products, air pollution fallout, and agricultural chemicals. However, they can be the primary source of pollution in many areas.[30]

When the Clean Water Act was passed, the goal of eliminating polluted discharge into surface water by 1985 was pledged. There was also an interim benchmark of making swimming and fishing safe in all surface waters. To meet these lofty goals, the EPA required pollution controls in both industrial plants and sewage treatment plants. However, the improvement was so slow that amendments were added to the Clean Water Act in 1981 and 1987. They first made funding available to local governments to develop projects to control municipal storm sewers.[31] Later, the Clean Water State Revolving Fund was established to improve municipal sewage treatment plants and systems among other water quality improvements. But the impact of these efforts on surface water quality was not readily apparent and there were local reports of it getting worse. In response, the US Congress established the National Water-Quality Assessment Project (NAWQA) in 1991.[32] This assessment was to determine how both surface and groundwater quality has changed across the nation and how it is expected to change in the future.

IMPACT OF THE CLEAN WATER ACT

The surface water portion of the NAWQA project investigated 633 stream and river sites between 1992 and 2012. This and other water quality studies determined that between 1972 and 2001, there had been a 12 percent increase in surface water bodies that were safe for fishing nationwide though only 19 percent of them were examined for this study.[33] However, by the end of the study, it was clear that the efforts were having positive results in many areas.[34] For example, there was a significant decrease in the amount of nitrogen and phosphorous in urban rivers after the Clean Water Act began.[35] This change is interpreted to be the result of improvements to wastewater systems. In contrast, surface water in agricultural areas saw minimal changes and actually increased in phosphorous as often as it decreased. In general, agricultural areas have had the least improvement and, in many cases, got worse. The nitrogen compound of ammonium, however, has decreased across all areas. This is an overall improvement because ammonium is toxic to many aquatic species. It can be from wastewater or air pollution fallout. Sulfur in surface water across all areas has also decreased as the result of decreased air pollution fallout.

Studies show that urban rivers have improved the most and across all contaminants studied. Surface waters from undeveloped and mixed areas have improved in most contaminant areas studied. In contrast, agricultural areas have degrading water quality in most contaminants. Regarding the urban rivers described, the Hackensack River, the lower Passaic River, and Berry's Creek are all EPA designated Superfund sites with high-budget remediation projects currently underway.[36] All of the industrial sites described are also Superfund sites and in various stages of remediation. All of the landfills are closed and are now just grassy knolls. The Hackensack Meadowlands is now a large, managed natural area and a bird sanctuary. The surface water still contains significant levels of pollution but it is much better than it was.

Chapter 7

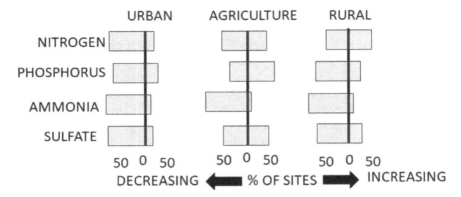

Figure 7.1.   Chart showing the relative change in pollutants in urban, rural and agricultural area rivers during the US Geological Survey national stream survey. The gray bars represent 100% of the stations. Bars shifted to the left (decreasing) show the percent of stations showing improvement in each pollutant. Bars shifted to the right (increasing) show the percent of stations showing worsening in each pollutant. Urban rivers show the most improvement.
Source: Adapted from Stets et al. (2020).

The Cuyahoga River is designated as an Area of Concern in the Great Lakes Water Quality Agreement.[37] The agreement is between the United States and Canada and it is repairing the damage done to the Great Lakes by all of the industrial and municipal pollution. The remediation of the Cuyahoga River includes reduction of eutrophication, toxic substances including PCBs, and heavy metals in river water and sediment. It also addresses bacteria from sewer discharge, and overdevelopment. As a result, the Cuyahoga River is now much cleaner and safer with neatly developed waterfronts locally and even a natural recreation area in the upper reaches.

Many other urban rivers in the United States are greatly improved. The Mississippi River in New Orleans, Louisiana, was severely polluted prior to the Clean Water Act.[38] It is reported that the bacteria levels are about 1 percent of what they were prior to 1980 and the levels of lead are now about 1,000 times less than they were in 1979. Even the Hudson River in New York, which is a 200-mile-long Superfund site in which the sediments are contaminated with PCBs, is greatly improved.[39] It is the largest Superfund site and so contaminated that no fish could be eaten from it and there were PCBs damaging the quality of nearby groundwater. Rivers in most cities in the United States and most industrialized nations are far cleaner than they were fifty years ago. This does not mean that there aren't pollution accidents on occasion. These can still be devastating but they are rare.

Surface water pollution also is not quickly eradicated. During the COVID-19 pandemic lockdown, most urban areas witnessed dramatic improvement in air quality. Primarily just by not using transportation, the air quality improved. This is not the case with surface waters. The pollutants can be sequestered in sediments within the bed or the banks of the water body where it can be slowly dissolved into the water or swept into it by floods or shifting flow patterns.[40] Pollutants in floodplain sediments can be regularly washed back into the water body during precipitation events. Biota can also remove sequestered pollutants from sediments and introduce them back into the water bodies. This is why the fish in the Passaic and Hackensack Rivers still have high enough levels of dioxin and mercury to make them inedible even though neither has been released into the water for more than a half century.

REMAINING SURFACE WATER POLLUTION

Citarum River

Water pollution may be well improved in much of the industrialized world but there are still surface water bodies that are being actively polluted at similar rates to the worst cases ever. One of these is the Citarum River in Indonesia, which has been called the "world's most polluted river." Even worse, its catchment area has a population of more than 40 million people so the impact of the pollution is a major public health threat.[41]

The Citarum River is in the Bandung region of Indonesia and it is the longest river in West Java. It is used for irrigation and drinking water, and has three dammed reservoirs with hydroelectric plants that generate 5 billion kWh/year of electricity. It provides 80 percent of the water used in Jakarta, the capital and major city of Indonesia. In addition to Jakarta, it is the source of drinking water for multiple cities along its course.[42] It also is the source for irrigating farms that supply rice as well as fishery and livestock resources. The Citarum River provides about 20 percent of the gross domestic product of Indonesia.

The river flows through a part of West Java inhabited by more than 18 million people, 11.3 million of whom live in the river valley. In 2018, there were more than 2,700 industrial plants in the Citarum River basin and about 1,000 of them were along the river and utilized river water for processing. Many of the factories produce textiles and garments and they employ a large number of the local residents. These factories and the residents along the river are causing extreme pollution of the Citarum River waters. Until very recently, only about 47 percent of the factories had any wastewater treatment facilities at all and the majority of these were substandard. The rest of the factories piped their unprocessed wastewater directly into the river. In addition to this massive industrial waste, unprocessed household, municipal, livestock, fishery, and agricultural waste was also dumped into the river in far greater amounts than could be naturally attenuated. Sediment in run-off from erosion of agricultural and deforested lands made the water muddy and threatened the river with siltation.

Where the river passes through urban areas, the water surface can be covered by so much waste that it cannot be seen. Plastic bottles, Styrofoam objects, plastic bags, wood, and other trash are so thick that a person can walk across the river on them in some cases. The garbage is from many of the small villages along the Citarum River. They have no public garbage collection or landfills so residents burn their trash or throw it into the river. There are also no sewer systems available for 95 percent of the residents, further reflecting the poverty of the area. Much of the sanitation is through crude septic tanks at each house or the raw sewage is dumped directly into the river.[43]

As a result, the water quality in the Citarum River ranged from unhealthy to dangerous at its peak of pollution and it is not much better now. The river water is black and full of chemicals and dyes from the textile plants. Lead in river water was measured in concentrations of more than 1,000 times the USEPA limit for drinking water. Some reports, however, estimate the lead at 25,000 times permissible levels. Testing found aluminum at three times the world surface water average, manganese at almost six times the average, and iron concentrations at three times average. Testing also found the pollutants arsenic, cadmium, chromium, mercury, nonylphenol, paranitrophenol, PCBs, pesticides, phtalates, sulphites, and tributylphosphate at dangerous levels.[44]

All of the raw sewage discharge increases the organic content in the river. River water has been found to contain over 5,000 times allowable fecal coliform levels in some locations. E. coli bacteria levels were also exceptionally elevated and biological oxygen demand (BOD) was nine times the maximum allowable levels in research studies.[45] It was found that almost two thirds of the biological oxygen demand in the river was from domestic and municipal activities and more than one third was from industrial and agricultural activities. This pollution has killed off about 60 percent of the fish in the river since 2008.

A large segment of the lower Citarum residents live in abject poverty with no choice but to depend on the river both for water and their livelihood regardless of the health danger. Residents scavenge floating trash for items they can use or sell. They eat fish from the river in spite of the danger and even drink the untreated polluted water. As a result, at least 60 percent of the residents living around the river suffer from recurring skin infections from contact with the polluted water. Many regularly suffer from intestinal distress and disease from ingesting the water and fish. Residents also suffer from chronic bronchitis, renal failure, dengue hemorrhagic fever, and a significantly enhanced incidence of tumors relative to the rest of the country.[46] Children suffer significant developmental delays. It has been estimated that more than 120,000 people become seriously ill and about 50,000 die annually as the result of river water pollution in Indonesia. The dead include at least 20,000 children per year.

This horrible situation may be a good illustration of the conditions in rivers in the United States and other industrialized countries prior to cleanup efforts. However, this pollution is not just the result of poor environmental practices in Indonesia. Most of the textile factories are producing products for people in the industrialized countries and they are owned or funded by major corporations. They choose to produce products on the Citarum because labor is cheap and environmental standards are much lower than in their home countries. Both factors reduce costs allowing much greater profits for the parent companies. In some respects, the pollution of American rivers has just been moved to other areas rather than being eliminated.[47]

There have been several efforts to end and remediate the pollution in the Citarum River.[48] The Asian Development Bank issued a $500 million loan to Indonesia in 2008 to clean the river. A river revitalization project began in 2011, at a cost of $4 billion over fifteen years. The project barely made a dent in the pollution. The president of Indonesia began a seven-year project in 2018 to clean the river to potable levels. He mobilized more than 7,000 soldiers to clean up key sections of the river.[49] However, the lack of coordination of responsible parties, the lack of sufficient funding, bribes from factory owners to avoid changes to their operations, and silting from soil erosion after deforestation made this effort ineffective as well. Internet notoriety, government enforcement, and foreign consultants are beginning to have a positive effect by reducing upstream erosion and through local pollution awareness and anti-plastics campaigns. However, even now, the Citarum River is still viewed as the most polluted in the world.

THE TAKEAWAY

Surface water pollution, especially in urban rivers, has caused severe public health crises to the point of epidemics throughout human history.[50] Urban rivers in industrialized nations were extremely polluted and a threat to public health in the past and they still are in many areas.[51] It took extensive public pressure to reverse this danger in the United States to the point of overriding a veto of a sitting president, no easy task. Industry also strongly opposed these regulations which would eat into their profit margins. However, the demand for environmental health and safety prevailed and many urban rivers, though not perfect, are much cleaner than they were in the past and they continue to improve.

The general public can help with this effort by participating in local river cleanups, but mainly by understanding that everything dropped to or dumped on the ground will likely wind up in a river. This means that trash should not be dropped on the ground and old chemicals should be properly disposed of. Application of lawn chemicals and salt on sidewalks should be minimized because precipitation and surface runoff carries them to surface water. All of these help the situation but government action needs to be taken, in some cases, so petitioning elected officials and voting for candidates who will help the cause is essential.

Chapter 8

Pollution Time Bombs

Industrial production often produces very toxic wastes. During the industrial age and especially in the 1940s through the 1960s, more toxic waste was produced than could be properly disposed of. Further, because environmental laws were so lax, companies did not have to responsibly dispose of this waste. As a result, huge amounts of toxic waste were dumped and/or buried wherever companies could get away with it. This left tremendous amounts of materials capable of causing extensive public health and environmental disasters all over the United States and most other countries with no oversight or protection. Once these toxic time bombs began to become recognized, public pressure to the point of civil unrest forced the government to take decisive and effective legislative action despite their apprehension and, in many cases, unwillingness to do so.

The American environmental movement accelerated through the 1960s, reaching a fevered pace by the end of the decade. As one result, the first Earth Day was held on April 22, 1970.[1] It was encouraged by one politician and organized by a handful of amateur activists, primarily a graduate student. Even with this paltry group, the event was attended by about 20 million Americans, which constituted 10 percent of the population of the United States at the time. Considering the number of people in favor of an issue who will actually attend an event, and especially a poorly organized event, this massive outpouring of support was an eye-opener for the American public, industry, and elected officials. This show of intent helped precipitate a cascade of legislation and directed efforts despite the reluctance of the government in many cases.

This public pressure on elected officials was so great that the US Congress passed the Clean Air Act of 1970 and established the Environmental Protection Agency to enforce the new stringent air quality standards going forward in spite of the industry resistance. The following year, the United Nations recognized Earth Day, expanding the environmental movement to global levels. In 1972, the US Congress passed the Clean Water Act, which established stringent enforceable standards for water contaminants going forward. The same year, all of the persistent organic pesticides named by Rachel Carson as dangerous were banned by the US Congress. The next year, they passed the Endangered Species Act and in 1974, they passed the Federal Insecticide, Fungicide, and Rodenticide Act.

LEGACY POLLUTION

All of this legislation restricted contaminants released to the environment and the restrictions on them became increasingly stringent. As a result, the public fervor to protect the environment began to subside. People began to develop a sense of confidence and security in how the environmental issues were being addressed. However, this turned out to be a false sense of security. Industry had disposed of massive amounts of very toxic chemicals in inappropriate locations and settings all over the country wherever they could get away with it. These were pollution or toxic time bombs that began to become apparent and publicized in the media in the middle and late 1970s. This sequestered legacy pollution was at least as dangerous as the pollution potentially released at the time. It was as eye-opening to the American public and government as the pollution issues of the 1960s. Addressing it dominated the environmental activism movement for the following decade or more and it became part of the environmental culture thereafter. It also led the US Congress to develop the Superfund program to address these dangers.[2]

Love Canal

Likely the most infamous of the toxic time bombs was at Love Canal, New York. Love Canal was a small town southeast of Niagara Falls. There are conflicting accounts of how Love Canal was developed. The first account is that the land was originally developed to be the site of a canal designed to bypass Niagara Falls by a man named Love in 1834.[3] He had even begun excavation of a trench 3,000 feet (914.4 m) long, 60 feet (18.3 m) wide, and 40 feet (12.2 m) deep by 1836 before the development company went bankrupt and the project was abandoned. In the second account, the timing and intent of the canal was that it was part of a proposed model city on Lake Ontario and dug in 1894.[4] In this version, the US Congress passed a law to preserve the falls in 1906 which ultimately scuttled the canal project. This appears to be the favored account. In either case, this excavation or hole was named Love Canal by local residents and it covered an area of nearly 16 acres (6.4 hectares). It soon filled with rainwater and runoff and was used as a neighborhood swimming hole by the local residents until the 1930s. It was then purchased by the Niagara Power and Development Corporation.[5]

The excavation was later progressively used as a municipal dump for nearby communities including the city of Niagara Falls. With time, chemical wastes from the local petrochemical industry were also dumped into the canal. In 1942, the Hooker Chemical Company, which produced industrial chemicals, fertilizers, and plastics, also began dumping their chemical waste in the canal. In order to have exclusive rights to disposal at the site, they purchased the canal in 1946. Between 1947 and 1952, Hooker Chemical dumped more than 43 million pounds (19,505 million kg) of toxic industrial chemicals in drums and directly into the excavation.[6] There are even rumors that Hooker Chemical allowed the United States Army and some military contractors to dump waste from experiments on weapons of chemical warfare and even nuclear waste. By 1952, the canal excavation was filled and Hooker installed a hard-packed, clay and ceramic cover over the area to isolate the waste and prevent infiltration by precipitation and surface water.

More than 200 different chemicals were dumped into the Love Canal excavation. Most were pesticides, and organic chemicals including benzene, benzene hexachloride, carbon tetrachloride, chlorobenzenes, chloroform, dioxin, hexachloro-cyclohexane (HCH), methylene chloride, phosphorous, polychlorinated biphenyls (PCB), trichloroethylene (TCE), and toluene among others. By the early 1950s, a large open field was hiding possibly one of the largest hazardous waste repositories in the world.[7]

The geology of Love Canal is actually reasonable for waste disposal.[8] There is a layer of glacial till consisting of silt, sand, and gravel in a matrix of dense clay covering the area. The thick clay makes the layer essentially impermeable to fluid migration, making it appropriate for waste disposal. The liquid chemicals and waste leachate sitting in the excavation would not be able to infiltrate the groundwater system through the clay. Even the underlying bedrock is relatively impermeable in case any liquid waste could leak through. It was therefore not unreasonable to use Love Canal as a waste repository under lax environmental laws of the time. Hooker Chemical installed a tight-fitting impermeable cap that was also appropriate to isolate the waste. However, maintaining the integrity of the cap was absolutely necessary to prevent pollution problems.

In the meantime, the post–World War II national population boom put pressure on many communities to expand housing and educational facilities. The Niagara Falls area grew quickly and experienced these pressures.[9] The city of Niagara Falls offered to purchase the Love Canal capped chemical dump from the Hooker Chemical Company. The initial reason was to build a new school but new housing construction was soon considered as well. When Hooker refused to sell the property, they were threatened to have it condemned by the state and seized through imminent domain. In 1953, Hooker Chemical gave into the pressure and sold Love Canal to the city of Niagara Falls for one dollar, the minimum amount for the sale to be legal. Hooker disclosed that there were hazardous chemicals buried there and warned against developing the site but especially against the penetration of the cap. Through the wording of the agreement, Hooker attempted to absolve themselves from all liability in the future use of the property.[10]

When construction of the public school began in 1954, several of the excavation pits unearthed drums of unknown chemical liquids. As a result, this initial project site was abandoned and a new nearby location for the school was chosen directly on top of the landfill. Excavation of that site to build the foundation breached and removed part of the protective cover or cap of the landfill that Hooker Chemical installed in spite of the warnings. This project was the 99th Street School which was opened in 1955 for 400 students.

This was not the only penetration of the clay cap. As the population of Love Canal and resulting pressure for more housing grew, the 93rd Street School was also built as were about 200 houses and a large park. In 1957, installation of sewers and water lines also breached the cap in numerous places as did the construction of a major expressway.[11] With all of the excavations and undermining of the cap, it began to crack in other areas. The families who purchased these new homes were not informed about the landfill full of chemical waste beneath their feet or the potential public health risks.

The breaching, excavations, and cracking of the clay cap and liner allowed infiltration of precipitation and surface water into the hazardous chemical landfill below. The impermeable clay-rich till encasing the chemical waste was also impermeable to the infiltrating water. As a result, the landfill began to also fill with this infiltrating water like a bathtub. The winter of 1962 was very wet and accelerated the filling of the landfill. All of the toxic chemicals and drums began to float up to the surface and into neighborhoods. By the 1960s, the residents of Love Canal began reporting noxious odors, unusual puddles of liquids in their yards, and noxious liquids seeping into their basements.[12]

The reports were widespread enough that Niagara Falls officials attributed the odors to nearby chemical and industrial plants thereby downplaying the mounting problem. In fact, in 1976, Love Canal was ranked the fourth-best area in Niagara Falls in terms of social well-being of the residents. At that point, there were 800 houses and 240 apartments in the town. However, the same year reporters tested toxic chemicals in sumps in a number of houses. Children began developing odd skin irritations after playing on ball fields and pets came home with skin lesions. Then the western New York blizzard of 1977 dumped three to four feet of snow on the area and its meltwaters filled the landfill to capacity and it began to overflow.[13]

In 1978, the situation gained full public attention. Liquid wastes and contaminated groundwater flowed into Love Canal neighborhoods and collected in puddles. On August 1, 1978, the lead, front-page article in the *New York Times* described the leachate flowing from the surface of the landfill into streams, sumps, and low-lying areas throughout the town including on the grounds of schools. This article succeeded and preceded several investigative reports by local newspapers on the pollutants emerging in the town. It also spurred television coverage of the pollution and widespread recognition of the dangerous situation at Love Canal.[14]

Meanwhile the conditions around town continued to deteriorate. Rusting and leaking metal drums of toxic chemical emerged through the soil in backyards of houses and even in local playgrounds, all pushed up by the rising water in the landfill. A swimming pool was even floated off of its foundation by the rising groundwater and mixed in liquid chemical waste. Noxious puddles of toxic liquid began appearing all over town in ditches and other low-lying areas.

Since 1975, the Love Canal Homeowners Association (LCHA) investigated health concerns of the residents. They documented high cancer rates and increased birth defects throughout the town. On August 2, 1978, the then president of LCHA briskly stepped up her activism after learning that the epilepsy, asthma, urinary tract infections, and low white blood cell count that her son was suffering from were the result of exposure to chemical waste. With the help of investigative reporters, the LCHA even identified the Hooker Chemical Company as the source of the chemical waste. However, Occidental Petroleum, the new owner of the company, denounced the findings. Elected officials from the city of Niagara Falls and state of New York also reacted slowly to the news and attempted to downplay the findings and accusations. Even the concerns of supportive elected officials were dismissed by the local government.[15]

However, media and public pressure exploded from all directions to the point that President Carter declared Love Canal to be an emergency area and approved emergency financial aid on August 7, 1978. This marked the first time in United States history that emergency aid was approved for an environmental problem. The U.S. Senate further recommended allocation of federal aid to address the extensive pollution. The then New York governor Hugh Carey developed legislation for the state of New York to purchase the two hundred worst-impacted Love Canal houses. Within one month, ninety-eight families were evacuated from the town and forty-six families were placed in temporary housing. After two months, 221 families were evacuated, and the elementary school was condemned and closed.

Unfortunately, the government officials underestimated the outrage of the residents. Those residents of Love Canal whose homes were not purchased

or scheduled to be were very upset about the health threat to their families and their exclusion from aid and benefits. Their property values had
already plummeted as a result of the pollution which consequently greatly
decreased their personal assets. This unrest kept growing but was ignored by
the officials. The tipping point came in May 1980 when the Environmental
Protection Agency (EPA) released blood test results of Love Canal residents
that demonstrated chromosomal damage as the result of exposure to the toxic
chemicals. This greatly increased their risk of developing cancer and reproductive problems. The residents were so enraged that they held EPA officials
hostage at gunpoint for up to five hours before the situation was defused. This
escalation of civil disobedience convinced President Carter to evacuate all
Love Canal families. As a result, nearly nine hundred homes were purchased
and the families were relocated.[16]

The remediation of Love Canal was extremely expensive and resulted in
the abandonment of the town.[17] It took approximately twenty-one years and
$400 million to complete. The toxic waste was not removed from the landfill
because it was too dangerous for workers and nearby communities during
excavation, processing, and transportation. Instead, a thick plastic cover
was installed and covered with a thick layer of clay over the sixteen-acre
(6.4-hectare) landfill and a surrounding forty-acre (16-hectare) buffer area
was added. Surface drainage and leachate collection systems were installed
around the landfill. Nearly 3 million gallons (11.4 million L) of contaminated surface and groundwater were recovered and treated per year. More
than 1,600 cubic yards of sediments that were exposed to the leachate were
excavated, treated, and shipped to a processing and disposal site. Hazardous
chemicals are still monitored at the site using groundwater monitoring wells.

The former residential area on and around the landfill was demolished
and is now vacant. An eight-foot (2.5-m) high, chain-link fence was installed
surrounding the area. Some 260 evacuated homes west and north of the site
were repaired and sold and ten new apartment buildings were constructed.
Recreational facilities were built right up to the fence. There are no street
signs for Love Canal remaining, and it is not on maps. Instead, the area was
renamed Black Creek Village.

Love Canal was likely the most infamous of the toxic time bombs but it
was far from the only one. Other polluted sites began surfacing around the
United States and were quickly reported by the media. Each of these attracted
public attention to the dangers of environmental pollution. Although aerially
restricted and only impacting a small number of people per incident, collectively, they fueled public outrage and compelled government action.

Chemical Control Corporation

Another of the infamous toxic time bombs that changed history was Chemical Control Corporation in Elizabeth, New Jersey. The company carted away hazardous waste from manufacturing plants for a price and safely disposed of it at a waste processing plant which they paid for. Chemical Control Corporation was incorporated in 1970 and operated in this capacity from 1970 to 1978.[18] The facility was situated on a two-acre (0.8 ha) lot in Elizabeth and collected a number of waste types such as acids, arsenic, bases, benzene, biological waste, compressed gas, cyanide, dioxin, explosive liquids and solids, mercury compounds, military nerve gas, PCBs, pesticides, radioactive waste, and toluene among others, many of which they did not have permits to accept.[19] Chemical Control Corporation aggressively collected waste at rates of up to 1,200 drums or barrels per month, many of which were delivered late at night to reduce the number of witnesses. They did not process this waste at nearly the same rate as they collected it. As a result, the small lot became overrun with both legal and illegal toxic waste. Barrels of fifty-five gallons (208 L) in volume covered the property end to end and stacked up to six high. There were also many aboveground storage tanks among other storage facilities, many of which were leaking and contaminating the entire site and nearby offsite areas. The company was regularly cited for improper storage of waste and for illegally dumping waste offsite around Elizabeth. After numerous violations, the state of New Jersey finally closed down the site in January 1979.

The New Jersey Department of Environmental Protection (NJDEP) dealt with this pollution nightmare over the next fifteen months.[20] They drained numerous hazardous liquids from seven above-ground storage tanks and disposed of them. They identified the chemicals in 10,000 of the more than 50,000 barrels of chemical waste in the warehouses and stacked around the property and disposed of it. New Jersey was forced to bear the financial responsibility of the work and had to remediate the site on its own. By April 1980, most of the liquids and solids on the surface had been removed. Most of the explosive liquids, radioactive waste, and infectious wastes were also removed even though the company was not licensed to collect them. The NJDEP was preparing additional charges against the company owners as a result.

However, at 10:54 p.m. on April 21, 1980, a suspicious fire was reported at the Elizabeth site.[21] It spread quickly from its point of origin in a warehouse that held more than 20,000 drums of toxic waste. Firefighters arrived at an inferno with multicolored flames shooting hundreds of feet into the air. Periodically, drums filled with toxic waste exploded, launching others hundreds of feet into the air. The fire took more than ten hours to contain and firefighters remained at the site for more than 185 hours. The firefighters not wearing proper protective gear were quickly overcome by fumes and more than sixty of them required hospitalization. Most of them experienced long-term symptoms, primarily involving the lungs and skin. Ten years later, twenty of these injured firefighters had developed cancer and other potentially terminal diseases.

Most residents of Elizabeth and many in nearby Staten Island, New York, were awakened by the explosions from the burning waste. The firefighters and emergency personnel labored to control the four-alarm blaze while officials from NJDEP and New York Police Emergency Control Board scrambled to assemble evacuation plans for Elizabeth and the New York City Borough of Staten Island. If the cloud drifted too far to the east, they would need to evacuate parts of Brooklyn as well which had five million residents at the time. Hundreds of thousands of people were being prepared to be removed from hospitals, nursing homes, apartments, and houses. However, the wind shifted direction and the cloud of toxic chemicals that had formed above the fire slowly drifted southeastward and out to sea.

The fire could have been much more devastating. NJDEP had already removed many of the more hazardous chemicals. Their absence reduced the toxicity of the smoke and ash. The fire was also so hot that it incinerated many chemicals, further reducing the smoke toxicity. However, smoke and ash covered fifteen square miles (38.9 km^2) in Elizabeth, Staten Island, and other nearby towns which forced Elizabeth schools to close. Residents were recommended not to leave their homes for several days. After that, they were advised to wash cars, lawn furniture, and any other outdoor items to remove contaminated ash. Within six months, the firefighting equipment was decontaminated.

The remaining drums and liquid chemicals were also removed within about six months.[22] Within eighteen months, the drums that had been dumped, launched, or fell into the Elizabeth River were also recovered. The warehouses and other structures on the property were demolished and removed. The sediments in the Elizabeth River that were contaminated by chemical-infused runoff from the facility and firefighting operations were dredged, removed, and decontaminated. It took more than five years to dispose of 187 gas cylinders of unknown contents and potential danger. The site was isolated from the Elizabeth River by a soil berm, and a chain-link fence isolated it from the public. As of 2019, the site was still not declared safe for use.

Newspapers reported that the owner of Chemical Control Corporation was a member of the Genovese organized crime family.[23] Further, the fire at the facility was a case of arson in retaliation for New Jersey's takeover of the property and the resulting loss of revenue. There were also reports that the fire was set to hide even worse infractions. These claims were reinforced by testimony from a reliable associate of the owner. As a result, criminal charges were filed against the owner of the company as well as one of the waste transporters who diverted toxic waste to the facility, instead of processing it. Both were convicted and sentenced to prison terms.

Valley of the Drums

The third toxic time bomb cited by the EPA as a cause for legislative action was the Valley of the Drums site.[24] The A. L. Taylor or Valley of the Drums site was another of the infamous pollution time bombs that attained national attention and helped spur legislative action. The twenty-three-acre site is located in Brooks, Kentucky, ten miles from Louisville and only 125 miles (201.2 km) from Tennessee where former vice president Al Gore grew up. The infamy of this site was part of the impetus for Gore to take an interest in environmental activism.

The site contains seventeen acres of woodland and grassland with a security fence enclosing another six-acre area.[25] The site is surrounded by wooded areas, private homes, the Wilson Creek, a golf course, and a few commercial businesses. In 1966 and 1967, A. L. Taylor started a small, steel drum recycling business on his own property. The operation was in a stream valley with high walls and steep slopes on one side and lower, gentle slopes on the other side. Taylor convinced local industrial paint and coating companies in Louisville, Kentucky, to send him empty and partly full drums of solvents, paints, and chemicals for disposal and recycling. Upon receipt, Taylor emptied the leftover chemicals into five large unlined pits he dug on his thirteen-acre (5.3-hectare) property. He further washed out the drums with detergent and organic solvents which were also dumped into the pits. He then returned the drums to the customers or sold them to other companies as reconditioned drums. When the pits were filled, they were set on fire to burn off the excess waste. Taylor did not obtain necessary permits to operate this waste disposal facility from the state of Kentucky so he was able to provide the service for very low cost. As a result, Taylor got a lot of business and operated until 1977.

The A. L. Taylor Drum Cleaning Service was brought to the attention of the Kentucky Department of Natural Resources and Environmental Protection Cabinet (KDNREPC) early in the operation of the disposal business.[26] In 1966, there was a report that a fire had been burning in a waste pit for a week, producing excessive smoke. KDNREPC ordered an end to the burning and soil from the hillslopes was used to cover the burned areas. However, Taylor continued to accept and dump waste in ever increasing amounts. Finally, in 1975, the KDNREPC began documenting the release of toxic chemicals at the site and initiated legal action against Taylor. This prosecution was never able to compel Taylor to comply and he died in 1977 still disputing the charges.

KDNREPC was overwhelmed by the extent of the site so requested assistance from the EPA in 1979 and they complied.[27] They identified 17,051 drums at the site, including 11,629 empty drums. Rusting and leaking drums were releasing contaminants directly into Wilson Creek. They documented more than 140 toxic chemicals including heavy metals, volatile organic compounds (VOCs), plastics, polycyclic aromatic hydrocarbons (PAHs), and polychlorinated biphenyls (PCBs). They were identified in groundwater, surface water, sediment, and soil, all of which posed public health threats. The EPA secured the leaking drums, built interception trenches to divert surface liquid waste, constructed a water treatment plant, and cataloged the drums to separate out those requiring special handling. The EPA turned the operation back over to KDNREPC, at least temporarily.

Fortunately, the threat from the toxic waste to public health and the environment was essentially just its presence. Surface runoff from the site entered Wilson Creek, which flowed into Pond Creek, and then into Salt River before it reached the Ohio River. By that time, it had been diluted with other surface water by a factor of a million to one making it well within regulatory limits. Groundwater in the area occurs in a shallow soil aquifer and a deeper separate aquifer. The shallow groundwater cannot infiltrate into deeper levels and simply flows toward Wilson Creek like the surface water. The flow is so slow that it was estimated that a plume would move fifty feet in twenty years. The deep groundwater aquifer is protected from surface infiltration by an impermeable layer and is therefore safe from contamination. Further, the groundwater flow rate in the deep aquifer is about five times the average of deep aquifers in the state which would quickly dilute any infiltrating contaminants. Therefore, Valley of the Drums was not as dangerous as the other two examples.

COMPREHENSIVE ENVIRONMENTAL RESPONSE, COMPENSATION, AND LIABILITY ACT—SUPERFUND

There were numerous other examples of toxic times bombs that emerged in the mid and late 1970s. Several that impacted human health were recounted in books and movies. Examples of these include *A Civil Action*, which was both a movie and a book about the public health impacts of toxic chemicals in public groundwater wells in Woburn, Massachusetts. A number of people contracted and died from cancer from this exposure, including several children. The movie showed the heart-wrenching impact of the loss of a child. Another famous case was in Hinkley, California, where a major utility company, Pacific Gas and Electric, contaminated the town and area with deadly hexavalent chromium. Erin Brockovich was the legal assistant in the court case about whom a movie was made. In this case, the company paid for the cleanup and restitution to the residents. However, in many of these situations, this was not the case.

The problem was the inability of the government to level criminal charges against the offenders or to force them to pay for these expensive remedial projects. In Love Canal, the state of New York paid for the first part of the relocation and remediation but several federal government infusions of funds helped cover other costs while still under New York State's oversight.[28] Occidental Petroleum Corporation which had bought Hooker Chemical also helped pay for the cleanup and relocation. The remediation of the Chemical Control Corporation site cost New Jersey $26 million (US) in 1980. Almost 80 percent of the cost was collected from the 200 companies that generated the waste and the state of New Jersey covered part of the cost.[29] Most of the companies had already paid to have it properly and legally disposed of by Chemical Control Corporation so wound up paying twice. Neither of these remedial actions could be completely paid for by the state or companies and there were only state funds available to address the Valley of the Drums pollution.[30] This lack of funding certainty was clearly not a workable situation.

The US federal government made several efforts to pass legislation to address toxic time bombs. The first action was in 1976 when the US Congress established the Resource Conservation and Recovery Act (RCRA) which allowed the EPA to control generation, transportation, and disposal of hazardous waste.[31] The same year, they enacted the Toxic Substances Control Act (TSCA), which allowed the EPA to protect public health and the environment by controlling toxic chemicals that pose risk of injury. However, neither of these provided the legislative power to deal with sites that were already profoundly polluted. Public pressure was still mounting on the government to deal with this very dangerous situation. At that point, the EPA estimated that there were between 32,000 and 50,000 legacy hazardous waste sites in the United States and between 1,200 and 2,000 of them could cause public health crises.

The breakthrough came in 1980 when President Carter signed the Comprehensive Environmental Response, Compensation, and Liability Act (CERCLA) into law. This law addresses the dangers of hazardous waste dumps that are abandoned or uncontrolled. It allowed development of a national system for emergency response, assignment of liability and responsibility and cleanup of hazardous waste sites. It also created the Hazardous Substance Response Trust Fund or "Superfund" which collected taxes, fines and penalties, and cost recoveries to pay for the cleanups.[32] The profoundly polluted sites were added to the National Priorities List and were termed "Superfund sites." Appropriately, New Jersey Congressman James Florio sponsored the bill. In addition to the Chemical Control Corporation incident another of the seminal incidents in the development of CERCLA was also in New Jersey. In 1977, a welder's torch ignited a chemical-waste treatment facility in Bridgeport, New Jersey. The explosion and fire left six dead and thirty-five hospitalized.

All but the Hinkley, California, site of those described became Superfund sites. The Valley of the Drums site was added to the National Priorities List in 1981 and the cost of cleanup was assumed by Superfund. It was remediated and removed from the list in 1996, considered completed. The Chemical Control Corporation site was also added to the National Priorities List in 1983 but was not quickly removed and still not removed by 2019. Love Canal was added to the list in 1983 and then removed in 2004 with no additional actions. The Woburn, Massachusetts, Wells G and H site was added in 1983 and received extensive remediation for decades. It was not removed from the list in the last review.

These were not the only sites added to the National Priorities List. The nominations and additions of hazardous sites filled the National Priorities List quickly.[33] There was a fixed amount of funding in the act under the Carter administration and it was expended in a short period of time. In addition, certain complications arose that were not addressed in the original legislation. In response, the 1986 Superfund Amendments and Reauthorization Act was proposed and passed in the US Congress and Senate. Then President Reagan was opposed to the bill and lobbied to fight it. However, public pressure was so great that even such a powerful president could not stop it. He signed the bill but made a public statement of opposition and reluctance.

THE TAKEAWAY

CERCLA has defused many toxic time bombs nationwide for more than forty years preventing numerous public health crises and protecting the environment. It has also made land in many key locations usable where it had previously been too dangerous. Removing the health dangers also improved the quality of neighborhoods and increased property values. Currently, nearly 1,900 Superfund sites have been assigned nationwide, a number of which have been remediated and removed from the National Priorities List. New sites are being designated every year.

This is another case of a serious pollution problem that was impacting public health and the environment. Once appropriate public pressure was brought on government officials, they enacted appropriate legislation and many public health and environmental disasters were averted. Considering how strong Reagan was as a president, to force him to sign a bill that he was opposed to, the power of a united public cannot be underestimated. It can be applied to any environmental issue, including climate change. As a result of this legislation, the public now has appropriate recourse for toxic waste that was disposed of improperly.

Chapter 9

Current Climate Change Efforts

Humans have clearly faced numerous serious environmental threats and disasters of their own making in the past. Many have been almost miraculously resolved by approaching them in a thoughtful and logical manner that employs the most appropriate measures. In many cases, there were champions of the efforts with no other motivation than to resolve the threat in the most expeditious and complete manner to protect human health and the environment. However, they also required the public to unite in demanding these best solutions and pressuring elected officials to enact them. It seems that this public outrage and a united front regardless of the opposition, underhanded professional and personal attacks, and cost is the formula required to successfully overcome an environmental threat in the industrialized world. In order to overcome the climate crisis or any other environmental threats, it is very likely that a similar solution will be required.

Viewing the situation with no political or ideological spin, it is clear that the climate crisis is a serious problem with many potentially dangerous to devastating outcomes. However, it is an environmental problem just like those described and overcome and with the right approach, it could be resolved in about a half dozen decades just as the rest were. What exactly constitutes the right approach is still being debated and suffering from political spin from groups with ulterior motives. This is resulting in confusion and the wasting of valuable time thereby allowing the situation to deteriorate even further. It is hard to predict what this deterioration will entail. The problem with predictions is that the outcomes, interactions, and impacts are constantly changing with many undesirable to dangerous surprises appearing each year. In plain terms, the situation is getting worse every year and it could very easily get out of control if action is not taken soon. What that really means depends on the predicting group and the time at which the prediction is made.

TOOLS TO FIGHT CLIMATE CHANGE

To put the challenge of replacing fossil fuel in perspective, it is necessary to see the starting point.[1] In 2022, 79 percent of the energy used in the United States was from fossil fuels and therefore emitted CO_2. Further, another 8 percent was from nuclear power plants and is therefore not renewable. If nuclear power plants are to be reduced or eliminated, this source cannot be relied on either. In addition, about 2.3 percent of the remaining renewable fuels are from corn and soy biofuels which are highly destructive to the environment and reduce food production as discussed below. This source should be eliminated. This means that we will have to replace about 90 percent of our 2022 energy production. This is a mammoth task.

There are about a dozen tools on the tool belt that could be used to deal with the climate crisis but the United States and several other industrialized countries have chosen to essentially utilize just four: solar panels, wind turbines, hydroelectric, and electric vehicles (EVs). There has been dabbling with several other tools but not widespread adoption. There is actually another practice that might be considered climate action but which was not developed for that purpose. When people fill up at a gas station, the words "may contain up to 10% ethanol" or something similar typically appear on the pump. This ethanol is produced by fermenting corn and distilling the by-product into this

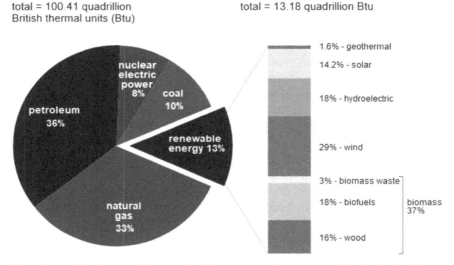

Figure 9.1. Pie chart and bar graph of the sources of energy used in the United States in 2022. Both nuclear electric power and renewable energy sources do not produce CO_2 but 79 percent of the energy usage in the United States emits CO_2.
Source: US Energy Information Administration of the US Department of Energy.

ethanol or alcohol. Driving through many Midwestern states, a driver passes by many tens of miles of corn fields. However, this corn is not used for food, it is used for ethanol and its growth, harvesting, and processing causes many environmental problems with few benefits.[2]

First, it requires a lot of fossil fuel to plow, plant, process, ferment, distill, and distribute the ethanol. The success in growing the corn is highly dependent on the weather. When the practice started, some years there was actually a loss of net energy. There have been many advancements in the process so that there is actually a gain of energy of about 30 percent in an average year.[3] Second, because the corn will not be consumed, it is all genetically modified (GMO) for maximum output. One of the major modifications is to make the corn impervious to glyphosate, the active ingredient in the weed killer Roundup. As a result, farmers can apply this dangerous chemical in as large quantities as they wish. The glyphosate spreads everywhere and kills off all of the native plant species, pest or beneficial. For example, the application of glyphosate has more than decimated the population of milkweed across the American Midwest as well as California.[4] The monarch butterfly depends on milkweed for reproduction. In the early nineteenth century, monarch butterflies were so plentiful that flocks of them could blot out the midday sun in the Mississippi River valley. As recently as the 1980s, there were more than 25 million migrating eastern monarch butterflies. Since the beginning of the application of glyphosate, the migrating eastern monarchs (east of the Rocky Mountains) have declined by about 80 percent and the migrating western monarchs (West Coast states) have declined by more than 99 percent. They are now endangered and on the verge of extinction. There are other factors in this decline such as pesticide usage which is also excessive with corn ethanol farming and habitat destruction but the overwhelming cause is the loss of milkweed. Third, this is prime farmland that could be producing food to feed the public. Considering the scarcity of food in some areas of the world, this misuse could be regarded as a huge waste of resources. The price of corn for food has doubled because so much land is used for corn ethanol. It has been estimated that the energy consumed in one twenty-five-gallon tank of ethanol fuel could feed a person for a year.[5] Fourth, the farming of the land actually increases the carbon footprint of the ethanol to the point that it is larger than fossil fuel gasoline.[6] This doesn't even take into account the massive use of fresh water, much of it from aquifers that are in crisis or pesticide impacts.[7] Soybeans grown for biodiesel pose the same kind of threat.

Growing corn for ethanol and soybeans for biodiesel should be discontinued immediately. It is only being held together by government price supports and requirements, not supply and demand.

There is nothing wrong with the four chosen climate crisis fighting tools and they should most certainly be continued. However, each both lacks capacity and has shortcomings that will never allow the climate change problem to be resolved just by using those four alone. They are not reliable or effective enough to meet this challenge either on their own or together. Other largely renewable tools must be used as well. In addition, none of them removes carbon dioxide from the atmosphere. Therefore, even if they were 100 percent effective and adoptable, the best they could do is to stop the deterioration of the atmosphere at the current levels. They can never improve the situation.

Only the tools of solar panels, hydroelectric plants, and wind turbines produce energy. The EVs use energy more efficiently than normal fossil fuel vehicles making them a conservation tool. A lot of electricity is still required to power them. Solar panels, hydroelectric plants, and wind turbines have the same major shortcoming: they are dependent on the weather. Solar panels are also dependent on the time of day and season.

Wind Turbines

Wind turbines are appearing all over the countryside and are even being built for offshore farms. They are being regarded as the solution to the climate crisis and even any possible energy crisis by many. They are certainly capable of producing impressive amounts of electricity with no carbon dioxide so are one viable solution to the climate crisis. Recent estimates are that they are now generating about 82,000 megawatts of energy per year, which is impressive.[8] In comparison, the United State consumes 3.9 trillion kilowatts of electricity per year so much more is needed.[9] Wind produces 29 percent of the renewable energy in the United States or 3.8 percent of the total. The locations of wind turbines and turbine farms, however, are restricted and should be. Urban and suburban areas are generally undesirable for wind turbines because the land is expensive and large enough areas to host them are few. In addition, the turbines require periodic maintenance, which can cause delays in a busy area. Although rare, they can also fail, causing danger to the public. Blades have broken and fallen to the ground and there is a threat of the mounting tower breaking or falling over as well. They are also not ideally installed in wooded natural areas. Installing, maintaining, and operating commercial turbines can be disruptive to natural ecosystems. It has been suggested that the spinning turbine blades kill numerous birds (140,000–500,000 per year),[10] bats (500,000 per year),[11, 12] and flying insects and certainly placing them nearer to natural areas increases the chances of this happening. This damage to bird populations is reported to be significant which is certainly a shortcoming of wind turbines, if true.

In addition, many land-based wind turbines can be observed sitting idle for long periods of time. It takes a wind speed of about eight miles per hour

(13 kph) to generate any electricity though some newer models can begin generating at five miles per hour (8 kph).[13] A continuous wind of eight miles per hour (13 kph) is fairly strong and rare in many areas. Further, they don't achieve full power output until the wind reaches thirty-one miles per hour (50 kph) which is a very strong wind even for gusts, much less continuous wind. As such, there are better and worse areas for wind turbines depending on the velocity and frequency of wind. At present, just four states, Texas, Oklahoma, Kansas, and Iowa, produce about 52 percent of the electricity generated by wind turbines in the United States, with the remaining 48 percent produced by the other forty-six states.[14] Wind turbines are also expensive, costing about $3–4 million for a commercial device with installation and about $45,000 per year for maintenance depending on the system. If there is not enough wind in an area, installation of a wind turbine will not be cost effective. This is another limiting factor.

On the other hand, wind turbines also have generation limits for the top speed of the wind as well. Most wind turbines have a cutout speed of fifty-five miles per hour (89 kph).[15] This means that the system shuts down and does not produce electricity if the wind speed exceeds 55 (89 kph). Offshore wind farms near Block Island, New York/Rhode Island, successfully shut down during a recent major nor'easter when the wind speed reached seventy miles per hour (113 kph) and successfully turned back on when speeds dropped back to fifty-five (89 kph). The problem is that if the turbine spins too fast it can overheat and the mechanical stress from the wind can cause it to fail. This means that there is an upper limit to how much electricity a wind turbine

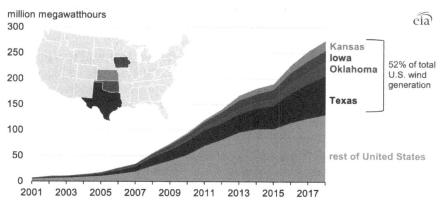

Figure 9.2.   Graph of the electrical energy generated by wind in the United States from 2001 to 2018 with the proportion generated by the four most productive states—Texas, Oklahoma, Kansas, and Iowa—separated. They produced 52 percent of wind-generated electricity in 2018. The inset map of the United States shows the locations of these four states.

Source: US Energy Information Administration of the US Department of Energy.

can produce. It also means that areas that are prone to high winds might not produce as much electricity as anticipated.

Offshore wind farms are now being viewed as saviors of green energy.[16] They produce about 40,000 megawatts per year in the United States at present. This is about half of the total wind electrical output. They have the potential to produce impressive amounts of electricity but they are restricted to coastal states with significant offshore areas. This means that the majority of states in the United States cannot benefit from them nor can landlocked countries. However, coastal states and countries can benefit greatly. The second problem is danger from powerful tropical cyclones including hurricanes, typhoons, and tropical storms.[17] The winds in these storms can greatly exceed fifty-five miles per hour (89 kph) so the production of electricity is minimal during them. It is also not clear how well the foundations and turbines can withstand the high wind and waves from such storms. If the turbines are damaged and destroyed every time a hurricane sweeps by or the foundations damaged in large storms, it will be a very expensive proposition.

Wind farms are planned for many areas along the United States coasts and offshore blocks of land have already been leased or are being leased.[18] Although to date there are only a handful of offshore turbines, the northeastern United States has multiple planned, very extensive offshore wind farms from Virginia to Maine. This area rarely has strong hurricanes but with climate change, that could change. In 1938, the Long Island Express or Great New England hurricane is estimated to have hit eastern Long Island with winds as high as 180 miles per hour (290 kph).[19] It then struck New England, killing about 700 people and making it the worst natural disaster in the region. It is very unlikely that wind turbines could withstand such powerful winds or the accompanying storm surge. The real danger is if coastal states convert their power grids to be reliant on the electrical output from their offshore windfarms. If a large hurricane damages or destroys these offshore windfarms, the electrical grids of several adjacent states could be reduced or even shut down. They would not be able to buy power from adjacent states and it would take years to rebuild the systems. However, hurricanes of this size are relatively rare in the Northeast so decades could go by before a scenario like this is encountered.

The Carolinas and Texas are also considering leasing offshore blocks for wind farms but they have a history of being struck by numerous very powerful hurricanes. This would place any offshore windfarm in jeopardy virtually every hurricane season. Indeed, any of the Gulf of Mexico states and countries and the southeastern states of the United States would face the same risk.

Wind farms in far northern coastal areas face icing in the winter. The sea spray and precipitation in these areas at temperatures below freezing could ice up the turbine blades. This would unbalance the load on rotation and make it uneven thereby reducing electrical output. The stresses caused by the extra weight in improper parts of blades could damage the axel hub and nacelle or generator as well. If the ice gets thick enough on the blades and hub, it could even prevent the turbine from spinning at all. For this reason, in very cold offshore areas wind turbines must be treated with deicing chemicals for part of the year to overcome these limitations.

Another location for wind farms is the northwest coast of the United States. This area does not have hurricanes but it does have earthquakes and can have tsunamis. Turbines in these areas will need to be constructed with the ability to withstand shaking which could increase the cost. Depending on the size of the tsunami, all turbines could be damaged or destroyed. Fortunately, powerful earthquakes and tsunamis are infrequent in that area.

A different danger that has been raised concerning the construction of offshore windfarms is the damage to the environment and especially the local ecosystem. The reports of whales being killed or injured by turbines have not been substantiated at this point but they are not the only marine life that could be impacted. Installing turbine foundations requires excavation and disruption of sediment where many marine species live. Cables must be installed from each turbine to a collection station which then has a larger cable to the shore. Cables are typically embedded in the seafloor, causing further disruption. Chemicals are used in construction and maintenance of turbines which may be released into the ocean also impacting marine life. It is also important not to build wind farms in the flight path of migratory waterfowl. Large birds would stand no chance against the quick slicing blades. All of these disruptions are minor for each turbine and can be considered inconsequential. The issue is that the blocks leased for wind farms are huge. For example, New Jersey has 537 square miles (1,391 km^2) of offshore in some form of leasing and there are an additional 927 square miles (2,401 km^2) of offshore being leased between New York and New Jersey. Many of the other northeastern states have similarly sized lease areas. If these huge areas are developed as wind farms, there could be ecological damage.[20]

There was a controversy about offshore windfarms with the US military. These large areas of impediment to ships and low-flying aircraft could impact training and any rapid deployment. The US military became concerned with the impact of the impediments but the controversy seems to have subsided. But it brings up the potential problem for any ships and low-flying aircraft. The farms and individual turbines may be large and spaced out enough not to be that problematic in clear weather but in storms, it could be another story. In rough conditions, ships and airplanes have difficulty maintaining their course. Having large stationary structures in the ocean could increase marine and aeronautical accidents. This could not only jeopardize the turbines but also make the oceans far less safe for large commercial ships and aircraft.

Considering that the New Jersey port is the second busiest in the Western Hemisphere, this is not a minor issue. The economic and human cost of losing a large off-course freighter would not be inconsequential.

Wind turbines and wind farms are certainly viable and desirable sources of green energy. They should continue to be developed where appropriate. However, their limitations in location, weather conditions, and potential ecological and transport damage makes them inappropriate to solve the climate crisis on their own. Optimistic projections on wind energy estimate that 35 percent of electricity will be produced by wind in the United States by 2050 (less than 10 percent of the total energy demand). This is a significant increase over the 14 percent of electricity produced by wind in 2022.[21] Wind can be one of the very effective tools used to address the electrical demand but it would be better if they were used in conjunction with other sources.

Solar Panels

Another major effort in green energy generation is solar panels.[22] Like wind, there is nothing negative about using solar panels as a source of electricity to help power the electrical grid of states and countries. They produce no carbon dioxide and directly convert sunlight into electricity through a physical/chemical process. They are ideally used on rooftops, over parking decks and on industrial and other buildings and structures. These areas are not being used for anything useful so can be easily put to use to help provide needed electricity. In 2022, solar panels produced 14.2 percent of the renewable energy in the United States or 1.85 percent of the total energy demand.[23]

Like wind turbines, the problem is that solar panels are at the mercy of the weather, time of day, and season. Even on cloudy days, they produce minimal electricity, if any. Therefore, they are useful in some areas like the southwestern United States which has abundant sunshine almost every day but not so good in places like the coastal northwestern United States or Great Britain among others which are commonly overcast or foggy. In northern or high-elevation areas where they can be covered in snow, the panels are heated to melt it but can still go long periods without producing electricity at all. Further, it is not clear how much snow weight they can support before being damaged or if shoveling a roof can damage them. Considering the cost of the panels and their installation, this is no small concern. The shade from trees will also reduce or eliminate electrical output. Therefore, homeowners cannot plant large trees or allow them to grow tall around their house. Considering the amount of carbon dioxide that trees convert to oxygen through photosynthesis, this restriction is a bit counterproductive. Trees also reduce home heating and cooling energy usage and cost by acting as insulators. This means that many broad wooded rural and even some suburban areas are inappropriate for solar panels. Communities in deep valleys or at the foot of mountains will also not receive enough direct sunlight to be economic. Even in polar and near-polar areas, the sunlight is at such a low angle of incidence that even under the best of conditions, electrical output from solar panels is low.[24]

In addition to weather, solar panels are greatly restricted by the time of day and time of year. Solar panels produce no electricity at night or even in twilight or at sunrise. However, in many cases power demand increases at night so this is a significant limitation. In the winter, days are very short, further limiting the amount of time that electricity can be generated. The low angle of incidence of the sunlight during this time of year means that the rate of electrical production is much lower than in the summer even with the same duration of sunlight.[25] On the other hand, solar panels are more efficient at lower temperatures. This means that in very hot areas, the expected high electric output is also limited. These limitations cannot be overcome which means that solar energy can never provide all of the needed electricity without major changes to the systems.

The other problem is that conventional chemical solar panels are not that efficient at converting sunlight into electricity. Most panels that are on houses today have efficiencies ranging from about 15 to 20 percent depending on age and quality.[26] Newer panels are slightly more efficient than older ones. This is because of advances in the bed material, the improvement in electrical loss (recombination) reduction and making the surface less reflective so the sunlight is used rather than being reflected away. Solar panels that contain high-cost, Interdigitated Back Contact (IBC) cells are made of high purity N-type silicon and yield efficiencies of 21 to 23 percent but the cost is much higher.[27] There are also concentrator solar cells that focus sunlight yielding higher output but they are also costlier. Solar panel efficiency will continue to slowly improve and produce higher output but not by much and they will need to be kept clean for it to be realized.

Like wind turbines, solar panels are great advances in carbon dioxide–free energy production. Their use should be expanded in all appropriate areas to help produce more green energy. However, trees should not be removed just to improve the sunlight on a roof. Removing them is counterproductive to battling climate change. Similarly, constructing solar panel fields over otherwise productive land is counterproductive for the same reason. Solar panels shading the soil reduces or eliminates plant growth on it so less carbon dioxide is being converted to oxygen. It is not an issue in the winter when plants are inactive but in the summer, covering productive soil is especially inappropriate in wetter areas closer to the equator. Dry areas can have large fields of solar panels without much concern.

Another type of solar energy is solar concentration systems, also known as solar-thermal concentrators. These systems involve setting up a field of mirrors that reflect the sunlight to a central electric producing facility. The mirrors concentrate sunlight and direct it to a black structure in the center. This structure heats up water in the container to the point that it boils and creates steam which turns a steam turbine and a generator that supplies electricity. These units generate electricity well but require large areas for the mirrors and the whole region heats up. They also suffer from the same shortcomings as solar panels of only generating during the day and under proper weather conditions. They are far less common than solar panels.

Optimistic estimates are that if effort is made, solar energy could provide 20 percent of the total electric generation by the United States by 2050.[28] In 2022, it provided 5 percent of the electricity (1.85 percent of the total).[29] Clearly, this is not enough and is unlikely to ever be enough even with all of the other currently used renewable sources.

Hydroelectric Power

Hydroelectric power is another renewable source of electricity that is widely produced and used. Basically, large rivers are dammed, forming large reservoirs with significant drops in water elevation across the dams. The water is released from the reservoir through a penstock or gate where it turns turbines to generate electricity. Hydroelectric power accounts for about 6.3 percent of the electricity produced in the United States (2.3 percent of the energy total) but it is about at capacity so it cannot be expected to help with the climate change issue any more than it already does.[30] The largest hydroelectric plant in the United States is the Grand Coulee Dam on the Columbia River in

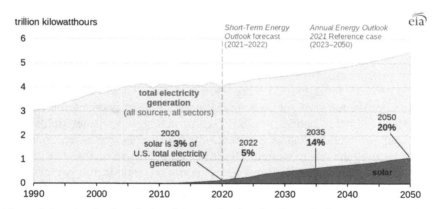

Figure 9.3.   Graph of total electricity production for the United States and that from solar energy from 1990 and projected to 2050. Solar energy production is now about 5 percent of the total and optimistically may constitute 20 percent of the total by 2050.
Source: US Energy Information Administration of the US Department of Energy.

Washington.[31] This plant generates about 6,800 megawatts per year which ranks fifth overall worldwide. The largest hydroelectric plant in the world is the Three Gorges Dam in China with a generating capacity of 22,500 megawatts and the second largest is the Itaipu Dam of Brazil with a generating capacity of 14,000 megawatts.

Hydroelectric plants require large rivers in areas of somewhat rugged topography to form the required elevation drop.[32] The rivers cannot be used for significant shipping transport because the dam blocks the course. As a result, there are few rivers and areas that are appropriate for plants. Further, the dam blocks sediment transport so the reservoir silts up disrupting the local ecology. In addition, the reservoir must be shut down and dredged periodically. The dams disrupt the migration and reproductive patterns of many species of fish in the rivers though many dams have been equipped with fish ladders to overcome this shortcoming. Finally, hydroelectric power depends on the weather like solar and wind though not always to the same degree. Droughts can limit electric output in the short term by reducing reservoir height and dam discharge. If there is excessive development and consequent increased freshwater usage in the watershed area, it can limit the hydroelectric energy permanently.

Electric Vehicles

The final major tool to address climate change in the United States as well as in several other industrialized countries is electric vehicles or EVs. Unlike wind turbines and solar panels, EVs don't produce electricity, they just use it more efficiently than most conventional personal and commercial vehicles. The development of EVs started with hybrid vehicles. These hybrids have both gasoline-powered engines and electric motors that can act in tandem to power the vehicle in the most efficient manner. In other hybrid cars, the electric motor boosts gasoline engine and propels the vehicle at low speed instead of a fully integrated system. The running of the engine and braking charge the batteries in hybrids. Both kinds of hybrid systems greatly increase the efficiency of the vehicle.

In contrast, EVs have only electric motors. The batteries must be charged at either a home or remote charging station and they are much bigger than in a hybrid vehicle. Considering that generating electricity is from fossil fuels, charging a battery with that electricity and then running it through a motor introduces a number of inefficiencies, it could be reasonably assumed that the efficiency of a hybrid vehicle should be much better than an EV. The difference is that an electric generation plant converts fuel to electricity far more efficiently than a car engine. Therefore, EVs can be about twice as efficient as hybrid vehicles which are typically about twice as efficient as regular gas-powered vehicles all depending upon the vehicles.[33]

The current shortcoming of EVs is that they have limited travel ranges on each charge. This is not an issue for most commuting distances in urban and suburban areas but it presents great problems for long-distance travel. Recharging takes several hours depending upon the vehicle which lengthens the duration of the trip. Fast charging or supercharging recharges the batteries in twenty to thirty minutes and overcomes this shortcoming but it shortens the life of the batteries so it is recommended not to be done regularly. Charging stations away from home additionally were difficult to find in the past but they have been greatly increased, reducing that shortcoming in most cases. Probably the most serious problem is that charging enough batteries to power every vehicle needed requires a lot of electricity. With still relatively few EVs on the road, there is not that much pressure on state electric grids. However, some states are passing regulations to require all new vehicles to be EVs after a certain date (typically 2035).[34] This will take all of the energy demands of personal transportation and add them to the already overloaded state energy grids. This increased demand will need to be factored into energy planning. New or greatly expanded sources of electricity will need to be planned.

EVs run on large lithium-ion batteries at present. Lithium is a very light element allowing the battery to store a lot of charge for the least amount of weight and size.[35] This is important for usage of the vehicle to be feasible. There are other elements required for automotive lithium-ion batteries such as cobalt, nickel, and manganese which make them subject to material shortages. Lithium has traditionally been a cheap abundant element with limited usage. However, once the EV rage began and government rebates and tax incentives were offered, demand skyrocketed. The price of lithium increased about sixfold from 2021 to 2022 and cobalt increased as well, which worried EV enthusiasts and the market.[36] Lithium exploration and mining operations expanded to meet this new demand. Fortunately, as battery technology improved and battery companies emerged and expanded, the price of lithium-ion batteries remained about the same through this period. As rebate and tax incentives expired or decreased, demand also decreased. These changes caused the price of lithium to drop sharply between 2022 and 2023 and cobalt followed suit. However, the International Energy Agency (IEA) still predicts a worldwide lithium shortage by 2025.[37] They further estimate that the world will require about 2 billion EVs by 2050 to reach net zero on carbon dioxide release. This is quite a tall order.

The proven worldwide lithium reserves are nowhere near large enough to replace every conventional vehicle with an EV much less reach 2 billion units by 2050.[38] The countries with the greatest lithium reserves in descending order are Chile, Australia, Argentina, and China before the United States. There is currently a frantic exploration effort to identify more reserves so the order and amounts may shift wildly over the coming decade. In addition, there is no anticipated bottleneck caused by planned shortages or reduced production by producing countries at this point. There are also new technologies emerging that will reduce the need for cobalt and other elements in battery production. These all predict good outcomes for EVs in the future. However, lithium-ion batteries have limited life spans and typically require replacement before the vehicle is worn out. Replacements are very expensive.[39] There is no system of lithium-ion battery recycling yet established so this will further exacerbate shortages and cause disposal problems. On the other hand, there is also current experimentation with developing sodium-ion batteries for charge storage but so far they are bulkier and heavier for the same amount of electrical storage. However, if these issues are resolved, storing enough charge to run vehicles and even power houses will not be issues.

One of the greatest limiting factors for EVs is the cost. Although prices are slowly decreasing as more companies are producing cheaper EVs, they are still very expensive. In 2022, EVs averaged about $18,000 more than the average vehicle price ($66,000 versus $48,000).[40] This may not be a problem for the rich or upper middle class but people who are struggling to pay their bills will have difficulty affording these new EVs at least until the price comes down. Further, at this point, buying a used EV may be expensive and not viable. It is estimated that EV batteries will last about eight to twelve years at good output depending upon usage which is a long period. However, replacing them is estimated to cost between $4,500 and $18,000. That means people who cannot afford a new EV could be buying cars that cannot perform well because of depleted batteries and they may be burdened with such an exorbitant cost that they cannot keep the vehicle. Conventional vehicles do not suffer from such issues. Until a fair system is designed so that people of lower economic status can be guaranteed the right to reasonably priced personal transportation, EVs will be an unreasonable and biased requirement.

The second overwhelming shortcoming of EVs is that it is necessary to charge them regularly. According to the US Census Bureau, 80 percent of Americans live in urban areas.[41] According to the United Nations, on the global level, 55 percent of people live in cities but it is projected to increase to 68 percent by 2050.[42] For wealthy urbanites, this is not an issue. In no time, most major parking garages will have charging stations in every space. However, most people with lower incomes must park on the street. There is no way that charging stations can be installed at every single parking space in every city. Even if they could, the cost of repairs from vandalism and theft would be staggering. Certainly, no one could expect the comical specter of multiple cobbled-together extension cords from high-rise apartments to cars parked on the street. Again, EVs are not reasonable for lower-income residents at this point. Even home charging has its limitations. Not only must the home electrical service be able to handle the demands of charging, the street and area must have enough transformers to handle the increased load. If every house is charging vehicles nightly, this amounts to a very large change in electrical demand and service. The rebuilding of the national electrical system is not trivial.

Fighting climate change cannot just be a boutique effort for the wealthy. If climate change is to be overcome, everyone must not only participate but also feel that their effort is substantial and valued. Many urban residents from impoverished areas are currently asking what they can do to help fight climate change. The answer cannot be nothing.

As mentioned, however, the real problem is generating enough additional electricity to charge the vehicles on a daily basis.[43] Many states can barely keep their power grids functioning properly even using fossil fuels and without a large number of EVs. California typically suffers from the highest number of power disruptions in the United States but in 2022, Texas had the worst record as well as its usual longest duration of power outage.[44] However, in 2022, the prolonged heat wave in California led to numerous outages and caused the state to revise its schedule of closing down nuclear and fossil fuel power plants. The problem is that California passed a bill to require all new vehicles sold in the state to be EVs by 2035.[45] A dozen other states are considering similar legislation. Considering the trouble generating enough electricity for the states without the EV law, it is unclear how they will generate enough electricity to charge all of the EVs. Further, most people charge their EVs overnight at home when there is no solar energy input. Hopefully, there are plans underway to address these shortcomings.

THE TAKEAWAY

The four current tools that the United States and several other countries have chosen to overcome this new environmental challenge of the climate crisis are wind, solar, hydroelectric, and EVs. The wind, hydroelectric, and solar tools generate electricity and the EV uses it more efficiently than our current methods. All of these are excellent tools in the climate battle and should continue to be developed.[46] The problem is that wind, hydroelectric, and solar energy depend on the weather which is unreliable, can never generate electricity to meet demand at the time it is needed and are location restricted so they cannot necessarily deliver power where it is needed either.[47] EVs depend on lithium and other elements which may not be in high enough abundance to meet the demand as the only source for batteries for the entire world. There is certainly not enough now. The power grids of many states and countries are already at their maximum output before the planned massive switch to EVs which will put an enormous increase in demand on them. Without a quick major expansion and overhaul, it is doubtful that they will be able to meet the increased demand. Finally, the resolution of climate change using only these four tools requires significant financial outlay on a personal level. This means that most of the population will be precluded from participating in the effort. It may also put undue financial hardship on the people and families who can least afford it. This is certainly not the way that the previously described environmental challenges were approached and resolved. There is nothing wrong with the four tools but there are many additional tools that can be used and are used around the world that can involve everyone, not put undue financial strain on any socioeconomic group, be reliable, not cause or be subject to shortages and allow the world to resolve, not just slow, the climate change crisis in a timely manner that avoids the coming catastrophic impacts. It is unclear why they are not being pursued.

Chapter 10

What Else Can Be Done?

Wind turbines and solar panels as energy generators and EVs as energy conservers are excellent tools to address climate change. They have enough limitations, however, that they will never be able to solve the crisis on their own. In addition to these three tools, there are many other powerful weapons against climate change that can involve all residents of the United States and the world. It will take a coordinated use of most of these tools and the cooperation and participation of the world to best attack the problem. Further, they will need to be employed relatively quickly to slow, stop, and reverse the progression before the damage becomes too great. If the world is to solve the climate change crisis the way we have defeated lead, DDT, and many other environmental challenges, a different approach is required.

There are several potential sources of clean energy that can solve all of our energy needs and the climate crisis at once. The best known of these is nuclear fusion, the same process that powers the sun. Researchers have been attempting to harness it as an energy source on a human scale for decades. Recently there have been some promising experimental developments but even if all experiments work better than expected and development of commercial electrical generators based on this work proceeds flawlessly, it will take at least three decades to even begin broad usage of the energy. There is little doubt that it will take longer than this which means intermediary generation sources and conservation methods are needed for the planet to survive the climate crisis.

GEOTHERMAL ENERGY

Currently, only 0.2 percent of the energy consumed in the United States is from geothermal sources (1.6 percent of renewables).[1] It produces no CO_2, is basically free and it is totally safe for the natural environment. It could easily provide ten to fifty times the current production levels depending upon the investment we are willing to make.

Hot Dry Geothermal

One very effective and currently underutilized tool against climate change is hot dry geothermal energy. Geothermal energy is the heat that is naturally emitted by the earth at all times and released to the surface and outer space if not used by humans as an energy source.[2] This heat is generated by gravity, radioactive decay, and stored energy from extraterrestrial impacts and the sun from within the earth. Geothermal energy is classified in two types, hot dry and cold wet. Hot dry geothermal energy means that high amounts of heat are released to the surface or near-surface from igneous activity in the earth. Magma must be shallow enough in the crust to raise the temperature to or above the boiling point of groundwater within a few hundred feet of the surface. This occurs in volcanic and geyser areas which sometimes coincide.

Basically, groundwater above the shallow magma is heated to above its boiling point.[3] It cannot boil because the pressure from overlying water, soil and rock prevents it. However, if vented or pumped to the surface, thereby releasing the pressure, the water will boil. When liquid water is converted to vapor, it takes up more than twenty-two times as much volume at room temperature and pressure. At higher temperatures, it takes up even greater volumes. If the conversion from liquid water to vapor is instantaneous, it produces an explosion. This is how a geyser works. If the explosion is controlled, it can be used to generate electricity by powering a steam turbine similar to how a steam locomotive works. Hot groundwater is pumped under pressure into the turbine where it decompresses and rapidly expands, driving the turbine and rotating the attached generator to produce electricity. The steam condenses to water after it exits the turbine and is pumped back underground through a well to replenish the groundwater system. These systems run twenty-four hours per day, seven days per week with output levels easily controlled to meet demand to the capacity of the system.

Why doesn't the whole world use hot dry geothermal energy? Because it is restricted to very specific areas and has an energy output limited by the magmatic systems that power it. Where there is a good supply of near surface magmatic heat, hot dry geothermal energy can be a clean and reliable alternative to fossil fuel. For example, Iceland has extensive volcanism all across the island yielding a high heat flow that has been used as a heat source since the island was discovered. Iceland currently generates about 25–30 percent of its electricity and heats 90 percent of its homes using hot dry geothermal energy.[4] Most of the electricity is produced from hydroelectric plants so that 99 percent of the country's electrical usage is from renewable sources. On a per capita basis, they are the world leader in hot dry geothermal power production and usage at 755 megawatts. On a global ranking, the United States is the top producer at 3,722 megawatts followed by Indonesia, the Philippines, Turkey, New Zealand, Mexico, Italy, and Kenya before Iceland.[5] However, there are numerous other areas that are capable of producing geothermal power and many currently producing areas that can increase their output.

Geothermal power in the United States can be produced primarily in the Northwest from California through Washington, around Yellowstone National Park, in Alaska and Hawaii. With deeper drilling, a large part of the western United States could produce hot dry geothermal power though it may not be economically feasible at this time. The entire "ring of fire" around the Pacific Ocean can produce hot dry geothermal power from shallow levels. This includes the entire west coast of Central and South America, Mexico, Japan, the Kamchatka Peninsula, the Philippines, and Borneo among others. There are geothermal resources in the countries of the East African Rift System from Mozambique to Ethiopia in addition to Kenya and in addition to the Cameroun volcanic line in West Africa. Many volcanic islands can also produce enough geothermal power to supply all their electrical needs. These include the Canary Islands, the Lesser Antilles, the Azores, Barren Island, some Greek and Italian islands, and many South Pacific islands. Potential geothermal resources are not ubiquitous in these areas but there are enough prospects to be a significant resource. It is estimated that global hot dry geothermal resources have the potential to produce 318 gigawatts of electricity per year and yet only a bit more than 15 gigawatts are currently being produced.[6] This is only about 5 percent of the global potential. Even obvious areas are operating below their potential. For example, Hawaii still relies on fossil fuel for up to 75 percent of its electricity with only 17.6 percent currently from hot dry geothermal sources where it could produce most of the islands' energy needs from the abundant geothermal resources.[7]

As described, all power sources have shortcomings and geothermal power is no exception. The geothermal groundwater can contain large amounts of dissolved minerals. This is why there are such impressive colorful mineral pools in Yellowstone National Park. As the groundwater is pumped up the intake wells, it can precipitate minerals in the wells and on all pipes and equipment that it passes through before it changes to steam. These mineral crusts can require significant amounts of time and effort to remove, requiring regular prolonged shutting down of the systems for servicing. The other major potential problem is that many of the geothermal resource areas are volcanically active. If the volcanoes erupt, they can destroy the geothermal power plants or make the area around the plant too dangerous to operate. For example, Hawaii was producing 31 percent of the island's electricity using geothermal power prior to the last eruption of the Kilauea volcano. Now it only produces 17.6 percent of the electricity because of the ensuing damage and danger.

Cold Wet Geothermal

Cold wet geothermal energy is far more abundant and available but it cannot produce electricity. Instead, it can be used to help cool and heat homes and buildings thereby reducing the need for fossil fuels and electricity.[8] It is a conservation measure rather than an energy production measure. The basic science of wet cold geothermal is that the temperature below about six feet (1.8 m) under the surface is a constant 60 degrees Fahrenheit (15.6 C) plus or minus depending upon location. By circulating water from the ground into a building, heat can be transferred between the two similar to a heat pump in a home HVAC system but with much better efficiency. In hot weather, 60 degrees Fahrenheit (15.6 C) groundwater is pumped into a circulation system in the house where it is used to cool the air in the house. No conventional air-conditioning is needed. In cold weather, the same groundwater pumped into the house heats the inside air to 60 degrees Fahrenheit (15.6 C) and a conventional HVAC source heats it the rest of the way to the desired temperature. This means that phenomenally less energy can be used in heating and cooling both on an individual and global scale which is estimated to account for about 25–30 percent of all energy usage though some sources place heating alone as high as 50 percent of all energy.

There are two ways to install cold wet geothermal energy.[9] In areas of relatively shallow groundwater and deep or soft bedrock, a shallow well can be drilled and groundwater is pumped into the building where heat can be exchanged with indoor air. Once the heat is transferred to or from the groundwater, it is pumped out of the building and into a second well where it reenters the local groundwater system. The well depth depends on the size of the building to be heated and cooled and the hydrogeology of the area. The wells in large buildings can be 300 feet (91.4 m) deep or more. In areas of deep water tables, sealed systems are required. Basically, a radiator is buried around or near the building. Water is then circulated through the buried pipes where it exchanges heat or cold and equilibrates to 60 degrees Fahrenheit (15.6 C). The water is pumped back into the exchange unit inside the building to cool or heat the inside air as needed.

Cold wet geothermal has limitations. First, a large enough area is needed for a groundwater field or wells for temperature equilibration. In urban areas, this could be difficult to impossible especially in older developed areas where there is no place to drill. Crystalline rocks can be very hard, causing drilling to be slow and expensive. In areas underlain by these rocks and especially where they are at or near the surface, geothermal systems could be exorbitantly expensive. Areas of permafrost are probably also not easily engineered to provide cold wet geothermal heating both as a result of cost and freezing of pipes. Retrofitting cold wet geothermal systems to existing buildings is much more costly than installing them in new buildings.

There are large areas where cold wet geothermal cooling could be well developed and save enormous amounts of energy, especially from fossil fuels. The entire southeastern and southern United States from South Carolina all the way to eastern and coastal Texas is very hot for a large part of the year and also has a shallow water table. There is enough groundwater in this area to install two-well systems in virtually all rural and suburban locations where bedrock is deep. The coastal plain from Long Island, New York, through North Carolina is also appropriate for two-well systems and can be used for both heat and cooling. Buildings in the southwestern United States require extensive air-conditioning for much of the year. The groundwater levels are generally too deep for two-well systems and the lack of water limits their usage but sealed cold wet geothermal systems are ideal. The rest of the United States is location dependent. Most of the Midwest is conducive for installation but the Rocky Mountains have a lot of bedrock exposure at the surface and may not be as easy to develop. On a global basis, any sandy coastal area, desert, or sediment-rich continental interior would also be appropriate. This accounts for huge amounts of the inhabited land surface including most of Europe, the Middle East, North Africa, much of West Africa and the East African Rift, Australia, most of Asia, and most of eastern South America.

The main shortcoming of all geothermal energy is the cost.[10] Hot dry geothermal power plants require a significant startup investment, usually by the government. There has been some investment by governments but even though they quickly pay for themselves, there has been general reluctance to establish them. Cold wet geothermal systems for individual homes are also quite expensive largely because not much effort or governmental encouragement has been made to develop inexpensive systems. Considering the amount of carbon-free energy that can be produced from these areas, it is unclear why this is the case. Certainly, the public pays a lot of money for some energy including huge tax breaks and government incentives and price supports to oil companies and especially to farmers growing corn for ethanol and other crops, both of which are highly polluting.

ALGAL BIOFUELS

The next tool that is underutilized but a potentially phenomenal source of energy that actually removes carbon dioxide from the atmosphere as a side benefit is algal biofuels. Humans have been using biofuels the longest of any energy source primarily by using wood and other vegetation in fires to stay warm. Some vegetation can produce huge amounts of fuel and have become mainstays for energy production in certain countries. Brazil produces a huge amount of sugarcane to use in biofuel generation.[11] Sugarcane has a very high sugar content which is fermented and distilled into ethanol fuel. It is such a good fuel source that, unlike corn, it always yields a net gain in energy even if the weather isn't perfect. In Brazil, gasoline for cars and other vehicles is 27 percent sugarcane ethanol and it accounts for nearly 20 percent of the nation's energy consumption. As a biofuel, it removes carbon dioxide from the atmosphere which reduces the world's carbon dioxide levels if the excess vegetation mass is sequestered rather than burned or processed. Unfortunately, however, this is typically not the case. Many other wet and warm countries and areas also produce sugarcane biofuel like Jamaica and the states of Florida and Hawaii in the United States.[12] Cassava is also an excellent source of bioethanol and possibly even better than sugarcane because it has more fermentable material. This is clearly a direction that has promise in battling climate change.

Biodiesel has even more potential as a fuel source than bioethanol. Biodiesel was investigated during the energy crisis of the 1970s and became the butt of jokes about environmentalists because old cooking oil was converted into fuel and used in cars. However, biodiesel is a potent transportable fuel source.[13] It is a liquid that produces only water and carbon dioxide when burned fully and it produces less CO_2 than fossil fuel diesel. There is also basically no need for dealing with dangerous impurities like sulfur. It produces a lot of energy per volume at about 38 megajoules per kilogram, about the same as fossil fuel diesel, compared with gasoline which produces 45. It can be used in any diesel engine at a mix of 25–50 percent or more biodiesel and engines can be built to run on 100 percent biodiesel. It can also be used as home heating oil with similar applications and outcomes.

The United States grows significant amounts of soybeans in the breadbasket of the Midwest specifically to be used for biodiesel because of their high fat content.[14] Together with Brazil, they produce 60 percent of the soybeans in the world and the largest industrial (nonfood) use of this crop is biodiesel. The problem is that the average yield for soybeans is seventy gallons of biodiesel per acre per year which is not good and can result in a net loss of energy because the farming, production, and transport can use more.[15] It also suffers if the weather is not ideal. Like corn, the biodiesel soybeans are also genetically modified to tolerate pesticides so there is great damage to the environment in farming them. However, the second generation biodiesel crops like oil palms and jatropha have much better yields. Oil palm can produce 600 gallons (2,271 I) of biodiesel per acre per year which always yields a net gain in energy. Sunflower seeds, rapeseed, and vegetables among others can also be used to produce biodiesel but at a lower rate of efficiency.

Biodiesel has been recognized internationally as a good alternative energy source and there are even agreements to increase the usage threefold by 2030. However, under the current usage trends, it will never happen. Biodiesel use in the United States peaked in 2016 at about 2.1 billion gallons (7.9 billion I) but dropped to about 1.7 billion gallons (6.4 billion I) by 2019. It is primarily used as an additive to petroleum-based diesel.[16] The problem is that there are no economic incentives to use it and, in some cases, tariffs to keep its usage in check.

The other problems with these land-based biofuels are the same as with corn ethanol. First, they require the use of prime farming land that could better be used to produce food.[17] With the worldwide shortages of food for humans, using it is not a moral decision. Second, they require the removal of trees and even forests that remove more carbon dioxide from the atmosphere than the crops that replace them. As part of this problem, they require extensive habitat destruction that quickly reduces local and global biodiversity. Third, they require heavy usage of fertilizers and pesticides that are eroded and carried in streams to the ocean impacting many species and environments along the way. The fertilizer causes surface water eutrophication and contributes to the worldwide dead zone epidemic that is destroying ocean productivity. Dead zones result from eutrophication of ocean waters which is a marked reduction in oxygen content due to rapid growth of algae and oxygen consuming bacteria. The first dead zone was identified in the late 1970s and now there are well over 400 dead zones worldwide. In addition, the fertilizers are lost to reuse and the worldwide phosphate reserves, a primary fertilizer component, are already dangerously low. Fourth, they are perhaps even more dependent on the weather than solar or wind energy. In some years, the energy from the fossil fuels required to farm and produce the biofuel exceeds the energy in the biofuels produced yielding a net loss. In other words, it would have been better just to have used the fossil fuels than to expend the time, effort, and money needed to produce biofuels. This also means that they are seasonal; biofuel crops can only be grown during a short growing season in most areas. Only one crop can be produced per season in most cases.

So why is biofuel even included in the alternative energy possibilities? The reason is that there is an alternative source of biofuels that makes them more attractive than petroleum and possibly even EVs.[18] That alternative is to use algae for biofuels rather than land-based plants.[19] This idea is not new. Algal biofuels were extensively investigated for commercial production during the energy crisis of the 1970s but were largely abandoned when the oil supplies increased and costs decreased in the 1980s. Many species of algae have very high lipid or fat contents making them ideal to produce biodiesel.

Second, they grow extremely fast. Duckweed on lakes can double its biomass in sixteen hours and in the lab they are faster.[20] Algae is similarly quick. This is orders of magnitude faster than any other biofuel crop. Third, they can be harvested and processed in less than two weeks. This means that there can be numerous harvests throughout the growing season which, for algae, is much longer than land plants because they can be grown right up to the onset of cold weather. Fourth, they take up far less space than land-based biofuel plants. Algae can produce from 15 to 300 times the biofuel output per unit of land area compared to land-based biofuel crops depending on the algae and conditions.[21] It has been calculated that algal biofuels could meet the energy needs of the entire United States using just 13 percent of our land area and faster-growing species have been developed since that calculation was made. Land crops would take three to five times as much space. Even without the climate crisis, algae is a good alternative fuel.

In terms of environmental conservation, algal biofuel generators are also good solutions.[22] Algae is 50 percent carbon by dry weight so it removes huge amounts of carbon dioxide from the atmosphere. For every gram of algae produced, an amazing 1.83 grams of carbon dioxide is removed from the atmosphere. It is by far the least expensive and most efficient method to remove carbon dioxide from the atmosphere. This, however, depends on what is done with the biomass remaining once the biodiesel is removed. The left-over material can be burned for fuel which would return the carbon dioxide to the air, it can be processed into usable biomaterial products, or it can be buried and sequestered thereby permanently reducing carbon dioxide in the atmosphere. Of all alternative energy forms, only biofuel provides this option.

Algae is grown in tanks of water, either open or closed, depending on the type of algae and end usage.[23] It also requires fertilizer, most of the time. Both of these are limiting factors. The difference with the other biofuels, however, is that because the water and algae are contained, any phosphorus in the fertilizer can be recovered and reused. None of the fertilizer escapes to the environment so it does not contribute to surface water eutrophication or dead zones in the ocean. Minimal pesticides are used in growing algae and even if they are applied, they do not enter the natural environment so there is no loss of insects or vegetation. Algae biofuel facilities can be built in the most convenient location where they will least impact the natural environment and therefore not contribute to habitat destruction.

Algal biofuel generators are limited to certain locations based on weather and climate.[24] Algae can only grow where the temperature is high enough for growth and reproduction. It must be above freezing and normally above about 59°F (15°C) for reasonable productivity. However, there are multiple species of algae that can be used to generate biodiesel and they cover a range of conditions. This gives algae a much wider geographic range than other biofuel crops.[25] In addition, algae can be genetically engineered to meet specific biofuel production needs within a certain range. However, tropical areas can grow algae the entire year. Sunlight is also required so generally overcast areas are less desirable as are deep valleys and other generally shaded areas. The other limitation is that algae grows in water and requires a lot of it. Dry areas can only use enclosed generators because the evaporation rate is so high and the supply of water is low. This limits the species and amount of production. Algae is also very sensitive to contamination so growing conditions must be tightly controlled or whole batches of algae can be ruined.

Like crude oil, biodiesel can be engineered in a refinery to fit any need.[26] Through distillation and reactions using catalysts, virtually any type of hydrocarbon fuel can be produced from gasoline to natural gas and even biohydrogen for fuel cells. This allows people to keep using their current appliances, automobiles, and equipment without having to incur the cost of replacing everything with expensive electric and lithium-ion devices. This further allows all people to participate in battling the climate crisis rather than just the wealthy. There are also many cases where electric/lithium-ion sources of power are inappropriate. Besides the problems in urban areas and supply issues as previously described, it would be very difficult to recharge construction or farming equipment every night. Military vehicles, ships, and airplanes also require transportable fuel without having to charge batteries. Remote areas present a similar challenge with regard to providing sufficient amounts of energy rather than a transportable fuel. There are many other examples from commercial ships to airplanes. The electric/lithium ion solution therefore cannot solve the climate crisis by itself under any circumstances and algal biofuel is an excellent solution to bridge its shortcomings.

Although there is nothing wrong with EVs and the push toward more clean electric power, it is inconceivable that the government has not made more of an effort to develop algal biodiesel and other algal fuel products as alternative energy.[27] It will take some investment to develop the farms and research to develop the best algal strains and processes. But the technology exists to implement this energy source and excellent tool in the climate crisis battle. It will take some policy shifts and new legislation to make this happen but nothing radical.

NUCLEAR ENERGY

The most controversial of the possible tools to solve the climate crisis is nuclear energy. Because of nuclear accidents like Chernobyl, Three Mile Island, and Fukushima and disaster movies like *The China Syndrome*, nuclear power plants have gotten a bad reputation. They frighten many people, and this fear is not unfounded. However, modern nuclear power plants are not your grandfather's power plants. They are much safer and more efficient than they were in the past. They supply sufficient reliable power whenever it is needed, there is a huge supply of nuclear fuel, and they don't produce any carbon dioxide. Perhaps if there was no climate crisis, they might not be considered but clearly we cannot produce enough reliable energy from solar and wind sources, especially if we plan to pursue the EV course of personal transportation. Several very environmentally conscious countries are now producing most of their energy from nuclear power. France, for example, provides as much as 72 percent of its energy from nuclear power.[28]

Modern nuclear power plants use steam turbines like coal- or gas-powered power plants.[29] They generate heat using radioactive fuel, most commonly uranium. To be a good fuel, the uranium must have a large amount of the isotope uranium-235 so most fuel is first enriched to maximize the content though some reactors can use fuel that is not enriched. Once enriched, uranium can be used as fuel in power plants for from three to five years. Once it is no longer effective as fuel, it is very radioactive from fission products so if it is disposed of, it must be done carefully and the storage area must be carefully selected. However, it can also be recycled into nuclear fuel for reuse in power generation or in several other applications. France does this now.

Once enriched, the uranium fuel is manufactured into ceramic fuel pellets.[30] These pellets produce the energy equivalent of about 150 gallons (568 L) of oil each or one metric ton of coal. They are installed into twelve-foot-long (3.7 m) metal fuel rods which are bundled into a fuel rod assembly of more than 200 rods. The reactor core contains a few hundred fuel rod assemblies depending upon the size of the plant, each containing thousands of uranium fuel pellets. This reactor core provides the power to the plant. The fuel rod assemblies are immersed in reactor pressure vessels filled with water which is heated and used to drive steam turbines.

Nuclear fission of the uranium in the fuel rods provides the heat to the reactor core. The uranium atoms in the fuel rods are bombarded with neutrons that split them into two smaller atoms plus energy and additional neutrons. Some of these released neutrons bombard the uranium atoms in other fuel rods, in turn causing them to undergo fission as well. This releases more neutrons which drive more fission reactions in other fuel rods, producing a chain reaction process. The nuclear fission reactions in this chain reaction releases heat to the surrounding pressurized water which raises its temperature to about 520 degrees F (271°C). The hot water is piped from the reactor and depressurized so it boils to steam that drives the turbines and attached generators that produce electricity.[31]

The nuclear reaction is carefully controlled to produce the energy needed while preventing a meltdown of the plant and the potential disaster it could generate. The reaction is controlled by installing control rods that absorb neutrons to keep the neutron flux at the desired amount. These control rods are interspersed with the nuclear fuel rods. The control rods are made of neutron-absorbing material like silver and boron and they are raised or lowered to speed or slow the neutron emission and nuclear reactions as needed. After the steam drives the turbine, it condenses to water in a cooling tower. The condensed water can then be reused in the reactor. The heat of the entire power plant is cooled and controlled in a separate system using water from nearby lakes, rivers, or the ocean as needed. For this reason, nuclear power plants must be located on bodies of water.

There were 439 commercial nuclear power plants in more than thirty countries worldwide in 2022.[32] The United States had the most nuclear plants at ninety-two. These commercial plants are light-water reactors which means that they use water both for cooling and to control neutron energy. There are two types of light-water reactors that both use pressurized water. One type uses the cooling water to heat separate, non-cooling water for steam and the other type are boiling water reactors which produce steam directly inside the reactor from cooling water.[33] More than 65 percent of the current nuclear power plants in the United States are the safer pressurized-water types. They produce about 20 percent of the electricity in the United States and about 50 percent of the clean energy.

Unlike conventional fossil fuel power plants, nuclear power plants do not produce greenhouse gas or add contaminants to the atmosphere. In addition, the fuel requirements and consumption of resources in nuclear power plants are far less than conventional fossil-fuel power plants. For example, a single uranium fuel pellet contains the same energy as 1.1 tons (1 mT) of coal. Nuclear reactors consume an average of 29.8 tons (27 mT) of fuel per year compared with a similar output coal power plant which consumes at least 2.8 million tons (2.5 million mT) of coal for the same amount of electrical production. This means that small nuclear power plants can produce a huge amount of power and produce small amounts of waste. The United States produces about 2,000 metric tons of nuclear power plant waste per year.

Unfortunately, at present, nuclear power plants in the United States store the high level nuclear waste of spent pellets on-site. However, this is not the only option. The relatively small country of France has the second-most nuclear power plants at fifty-six.[34] They are using and developing fast-breeder nuclear reactors and they recycle the spent nuclear fuel. This recycling of spent fuel generates more usable nuclear fuel than they consume. Once these reactors are perfected they will use less than 1 percent of the uranium currently used in reactors. Recycling the waste for additional energy means that there would be very little disposable waste from power plants. The problem is that fast-breeder reactors and processing the nuclear waste are currently very expensive. That and laws that prohibit it is the reason it is not utilized in the United States. However, it would greatly extend the supply of uranium for power generation. It is estimated that there is about ninety years' worth of identified uranium reserves if it supplied all of the energy for the planet.[35] At the current level of usage and with the currently used technology, it would last 230 years. Recycling and use of fast breeder reactors could make our current uranium reserves last thousands of years (possibly 30,000 years by some estimates) at current energy demands. More minable uranium undoubtedly exists, so these are conservative estimates. There has been less uranium exploration in recent years with the lower demand. In addition, uranium can be extracted from seawater, if needed, which would greatly increase the amount of power available with no carbon dioxide or any other chemical pollutants produced. This source could power every EV in the world if that was the chosen option.

The final problem with nuclear power plants is the safety concern. First, the nuclear fuel used in a power plant cannot detonate in a nuclear explosion. The fuel is nowhere near weapons grade so it is not an issue. However, concern about enriching it into weapons grade material cannot be overlooked. Therefore, they should only be built in areas with stable governments. Second, the accidents that happened in the past, like the 1986 Chernobyl disaster, were because the reactors operated on different, less stable systems.[36] The Chernobyl reactor was also poorly maintained and understaffed which encouraged human error, the major factor in the accident. Modern nuclear power plants do not carry the same risks and are much better maintained. The most recent accident was the 2011 Fukushima incident, which was significant. It showed that building nuclear power plants in natural disaster-prone areas should be avoided. It is important to note, however, that the reactor survived a magnitude 9 earthquake, one of the biggest ever. It was just not prepared for a tsunami which caused the accident.

Modern and especially advanced nuclear power plants with fast breeder reactors should be considered as a tool to address the climate crisis. France is a very environmentally conscious country and depends on nuclear power for about 70 percent of its energy needs.[37] They export electricity to several neighboring countries that are abandoning nuclear power and experiencing shortages as a result. However, nuclear power plants, like all of the other climate crisis tools, have and should have geographic restrictions. The current modern nuclear power plants require a lot of water so they should not be sited in arid areas. New, advanced plants that use other cooling methods are being researched and planned to be widely available by 2030. Although they are very safe, they should not be sited in heavily urbanized areas. The reasons are the cost of land, lack of space to handle waste, risk to so many human lives if there was a problem, and that they could be more likely to be a terrorist target. Finally, they should avoid areas with a significant threat of natural disasters such as near an active volcano, in an earthquake-prone area, in an area of frequent landslides, a tsunami-prone coastal area, or in a strong hurricane or typhoon-prone area. If a nuclear plant was struck by a strong tornado, there could be an accident as well but these storms are small and rare so the chances of a direct hit are much less than the other risks.

This may sound very restrictive but it means that a corridor from New Hampshire to Minnesota in the United States where there are many lakes for cooling water could host numerous nuclear power plants. They would just need to avoid urban areas. South of this corridor, power plants could be sited along rivers but they should probably not be in tornado country unless they can be engineered to withstand a direct hit from a strong tornado. Some near-coastal areas are possible sites as well, as long as they are safe from large waves. These areas might be up estuaries or in protected bays most likely in the Northeast, avoiding the Southeast and Gulf Coast where there can be large hurricanes. The West Coast is seismically active and the Northwest is volcanically active so should be avoided. The southwestern United States is too dry, having very few sources of surface water for cooling. The operational area, however, could easily supplement the renewable energy sources like wind, solar, and geothermal to provide an uninterrupted, reliable source of power to facilitate any needs, even the widespread expansion of EV use. The trend is toward small modular plants (SMPs) which can provide a lot of power locally as needed on a small footprint.[38] These have a much broader geographic range than large plants because they use less resources and space and have less of a local impact while still providing needed power.

CARBON CAPTURE AND SEQUESTRATION (CCS)

There are some new and emerging tools against climate change as a result of government intervention that are in various stages of development and implementation. One currently active tool that actually removes carbon dioxide from industrial emissions is carbon capture and sequestration (CCS).[39] This tool captures carbon dioxide at the industrial exhaust source through filtering or scrubbing which is then either used for chemical processes or is permanently disposed of. CCS is especially useful for the many coal- and gas-fired power plants which account for about 40 percent of total carbon dioxide emissions. If widely adopted, this effort would remove carbon dioxide from all industrial sources so it would not contribute to climate change. The effort is almost completely driven by government investment and tax incentives for the polluting industries at this point and not widely practiced. The idea is that CCS would save major overhauling of industrial infrastructure to fight climate change and it even has some minor industrial applications.

Carbon dioxide capture begins with diverting and cooling industrial exhaust gases which can have as much as 25 percent carbon dioxide. The cooled gases are then passed through a solution primarily composed of amines which strip out the carbon dioxide. The amine-carbon dioxide solution is then heated in a regenerator until the nearly pure (85–90 percent) carbon dioxide is boiled off. The amine is then reused and the carbon dioxide is used or sequestered. Some chemical processes require pure carbon dioxide and it can be used to enhance productivity of sluggish oil fields. Some of the recovered carbon dioxide can be sold for these purposes but the rest must be sequestered where it will not reenter the atmosphere. It is compressed and transported as a liquid, preferably by pipeline. The plan is to pump or inject the carbon dioxide down old oil and gas wells a mile or more deep where it can be permanently stored. The carbon dioxide is a liquid at these high, deep underground pressures and it is stored in the same geological features, called traps, that previously stored oil and gas. They provide an immense and safe storage capacity for carbon dioxide.

This technology can dramatically reduce carbon dioxide emissions.[40] For example, a 500-megawatt coal-fired power plant normally emits about 3 million tons of carbon dioxide per year but would be reduced to 30,000 tons with CCS technology. This is equivalent to electrical-related emissions for more than 300,000 homes per year. Overall, CCS is now storing about 45 million tons of carbon dioxide per year in the United States all from large stationary sources like power plants and cement, steel, and chemical industrial plants. In comparison, approximately 68 million tons of carbon dioxide is injected into the oil fields for enhanced petroleum recovery in the United States so it is still in its infancy.

The first CCS system in the United States was installed at the Petra Nova-W.A. Parish generating plant in Texas in 2017 but it was shutdown in 2020 when oil demand and production decreased as a result of the COVID-19 pandemic.[41] Carbon dioxide demand also decreased and there was no way to get rid of it. The operation was never restarted. By 2021, however, there were twenty-four commercial CCS operations worldwide and twelve of them were in the United States.

The reason that this technology is being developed is because of government incentives rather than economic demand.[42] Since 2010, the US Congress has been providing $2.7 billion per year for CCS projects. However, the Infrastructure Investment and Jobs Act of 2022 provided $8.5 billion for CCS, including funds to construct new carbon capture facilities at power and industrial plants. It also provided $3.6 billion for facilities to remove carbon dioxide from ambient air which is much more difficult and expensive. The

US Congress has also incentivized development of CCS projects through the reduction of corporate income taxes governed by the Internal Revenue Code Section 45Q since 2008. The tax breaks were slowly increased until 2022 when they were greatly increased through the Inflation Reduction Act of 2022. This should greatly increase CCS projects and activity.

THE TAKEAWAY

These are just four of the tools on the tool belt that can be used to fight the climate crisis. It is clear that if these four were instituted to any degree in addition to several of the current practices, the climate crisis would be quickly be under control and even begin to reverse. The big hurdle to instituting many of these tools is the cost. However, the United States already invests huge amounts of money into energy projects. There are tax breaks for oil companies and US Congressional subsidies to Midwestern farmers to grow corn and soybeans for biofuels among other support.[43] If these funds were redirected to include the more climate friendly efforts described, the cost would be defrayed to some degree.[44] However, conservation efforts on a global level will also help defeat the climate crisis.

Even small conservation efforts can make a difference. Consider that it is estimated that the United States uses 27,000 trees per day just to make toilet paper and most of it destroys Canadian arboreal forests.[45] Imagine the carbon dioxide removing and oxygen producing power of all of these trees. No one is suggesting that we should stop using toilet paper but it illustrates that even small things that we take for granted amount to a huge impact with the nearly 8 billion contributing people on the planet. Widespread usage of recycled paper or fast-growing sources like bamboo or hemp for toilet paper or bidets as an alternative would greatly help this situation without reducing quality of life.

With this in mind, any efforts to reduce carbon dioxide emissions are beneficial. This can be reusing and recycling more products. The more recycled paper that is used, the fewer trees will need to be harvested. Even when glass bottles, aluminum cans, and plastic bottles are recycled, it requires energy to process them into something else. This energy typically involves the release of carbon dioxide. Therefore, reducing single-use containers should be practiced whenever possible. People love short, green lawns. However, trees take in much more carbon dioxide than grass. In addition, there is a new fashion of planting wild vegetation in place of lawns which also remove more carbon dioxide. Not using as many paper towels or wasting paper or cardboard of any type also reduces the number of trees harvested which will also have an impact on the climate crisis. Once these apparently minor conservation practices become widespread, they will have a large impact on the climate. Therefore, everyone can participate in battling the climate crisis.

Chapter 11

It Isn't Just Climate Change

The climate change crisis has expanded to essentially fill the entire bandwidth of environmental concerns and activism not only in the United States but also in most of the world. Governments are enacting laws to address the situation and it is seen daily on the news and in social media. It is fantastic that an environmental issue has become so prominent in the twenty-first century and that people are willing to take action. On the other hand, it is giving many people the impression that if they buy an EV and possibly get solar panels on their roof, humans will be safe from environmental disasters. This could not be farther from the truth. There are issues of ongoing environmental degradation that could threaten the human race long before climate change. This chapter will describe three of the direst environmental threats that humans face: freshwater shortages, the loss of pollinators, and ocean dead zones through overfertilization. All three are occurring at overwhelming rates and will eventually cause severe food shortages. The good thing is that everyone can take steps in their everyday life to help address them.

FRESHWATER SHORTAGES

Water is necessary for life on earth and for terrestrial life, much of it must be freshwater. The problem is that humans use inordinate amounts of freshwater for consumption, washing, agriculture, transportation, and recreation. Any even moderate reduction in the freshwater supply could be disastrous. Between 1876 and 1878, central China experienced a devastating drought that was responsible for the deaths of between 9 and 13 million people.[1] This normally moderate rainfall area basically had no precipitation for a year, and precipitation was well below average for a second year. The lack of water devastated the crops and a famine ensued, compelling people to do whatever they could to find food. There was not a blade of grass as far as the eye could see in this usually lush area. All of the trees were stripped of bark. There are reports of people selling their children for meals. Some people reportedly ate dirt for sustenance until it killed them. Whole villages are said to have committed suicide rather than slowly starve to death. Gangs roamed the countryside attacking people, not for their money, but to kill and eat them. There are even stories of people digging up recently buried corpses to eat them just to stay alive.

With the stockpiles of food and global trade, even if there are local extreme droughts, there are sufficient food and water resource and distribution systems available not to have a repeat of the central China disaster in all but the worst cases. But it illustrates how important fresh water is to maintaining a reasonable human society. If freshwater resources become too scarce, this type of horrific situation could become commonplace. The problem is that human society is headed in that direction in many areas and unless it is addressed quickly, there could be a global disaster.[2] Freshwater shortages occur in both surface water and groundwater supplies and they are currently occurring in areas where water resources were not an issue in the past.

Human water withdrawal from natural sources and subsequent usage on a global level has skyrocketed since 1900.[3] It grew exceptionally fast between 1950 and 2000, increasing by nearly fourfold in just fifty years. The growth of withdrawal has slowed in recent years but the current rate is still at peak levels and unsustainable. This is a global average rate but locally, the withdrawal

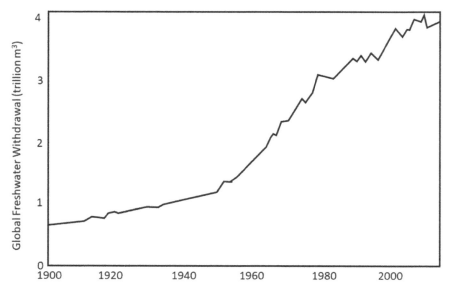

Figure 11.1.   Graph of annual global freshwater withdrawal from 1900 to 2017.
Source: Adapted from Ritchie and Rozer (2017).

rate exceeds the recharge rate by far greater rates. This situation quickly produces serious shortages to disasters.

Aral Sea Disaster

There are many areas of the world where fresh surface water resources are dwindling at an alarming rate but none are as alarming as the collapse of the Aral Sea in Kazakhstan and Uzbekistan. This collapse is considered by many to be among the earth's worst environmental disasters.[4] The Aral Sea is an inland lake that was named a sea because of its high salt content. Before the collapse, it was the fourth largest lake in the world. However, poor agricultural practices and overuse caused it to begin shrinking in the 1960s and by 1997, only 10 percent of the original surface water supply remained in four small bodies.

The Aral Sea supported thriving human and animal populations in an arid region. The lake is fed by several sources including the region's two major rivers, the Syr Darya and the Amu Darya, which are fed by snowmelt and precipitation in distant mountains. The problems originated in 1900, when a large area (7.4 million acres [3 million ha]) around the lake was irrigated for agriculture using Aral Sea water. The construction of irrigation canals for this agriculture began in the 1930s but were greatly expanded in the 1960s. Most of the canals were poorly constructed, leaking water into the soil and allowing excessive evaporation of the water they carried. The largest of these, the Qaraqum Canal, lost about 75 percent of its water as a result of poor construction. Regardless of these losses, until 1960, the massive removal of water was still balanced by the input from rivers, precipitation, and springs.

The situation changed in 1960 when the irrigated land was expanded by 68 percent (to 12.4 million acres [5 million ha]) and it continued to expand annually. By 1980, the irrigated land had increased by nearly 118 percent relative to 1960 (16.1 million acres [6.5 million ha]). The amount of water needed from the Aral Sea for agriculture was far more than could be replenished by the rivers. The irrigation canals wasted even more of the water during transportation from the lake to the fields. Estimates are that only 12 percent of main irrigation canals, 28 percent of irrigation channels between farms, and 21 percent of within-farm channels had liners to prevent infiltration of water into the dry soil below them. By 1987, the irrigated area had increased to about two and a half times its original size in 1960 (18.8 million acres [7.6 million ha]).

This massive irrigation project turned the surrounding arid land into an agricultural oasis but it devastated the Aral Sea. Between 1961 and 1970, the lake level declined about 7.9 inches (20 cm) annually. However, in the 1970s, the rate increased to about 20–24 inches (50–60 cm) per year and in the 1980s, it reached about 31–35 inches (80–90 cm) per year which radically dropped the lake level. This decline in the water level reflected the doubling of fresh water being directly diverted from the two rivers for irrigation between 1960 and 2000 so it never reached the Aral Sea. The lake level had been 174 feet (53 m) above sea level before irrigation. By 2010, it was 88.5 feet (27 m), a reduction in level by 85.5 feet (26.1 m) or almost half. The volume and area of the Aral Sea also reduced with reduction in lake level. From 1960 to 1998, the volume was reduced by 80 percent and the surface area was reduced by 60 percent. The lake split into two bodies in 1987. In 2003, the southern body of these two further split into eastern and western lakes and later it split into four lakes.

This disastrous drop in lake level and volume alarmed the local government.[5] They attempted to stop and reverse this trend in 2005 by building Dike Kokaral dam on the northern body of the Aral Sea. This effort appeared to have worked as by 2008, this lake level had risen by an impressive thirty-nine feet (12 m). All of the other bodies, however, continued to shrink. By 2009, the southeastern lake had completely disappeared. The southwestern lake had shrunk to a thin band along the former southern lake's west edge. By August 2014, the eastern body of the Aral Sea had completely dried out and became the current Aralkum Desert.

The extreme evaporation in this dry region and the diversion of incoming fresh water from the rivers to irrigation caused the salinity of the lake water to increase about tenfold from 1960 to 1987 in the remaining southern Aral Sea and basin. It reached more than twice the salinity of average seawater by this time. By 1990, the salinity in the remaining lake had increased to ten times that of seawater. It was essentially not usable for anything. This skyrocketing salinity of the lake water was further contaminated with runoff loaded with fertilizer and pesticide from the agricultural fields. As the evaporation progressed, the concentrations of these pollutants rose drastically in remaining water and they were concentrated in the dry lake beds of salt and sediment.

Prior to 1960, the Aral Sea had more than 24 species of fish, over 200 macroinvertebrate species, and 180 land animal species living along the shores and environs. As the lake water evaporated and the salinity increased, the fish were slowly eradicated and animal communities that depended on them disappeared. Only about thirty macroinvertebrates remain, as well as a few dozen land animals. The increasingly saline lake water was also polluted with pesticides and fertilizer which dried to contaminated dust in the exposed lakebed. This dust was blown aloft by the strong wind where it became a public health hazard. This dust settled onto agricultural fields where it degraded the soil, reducing the productivity. A 2001 United Nations study found that 46 percent of the irrigated agricultural lands were damaged by the extreme salinity of the lake. This is an increase from 42 percent damaged lands in 1995 and

38 percent in 1982. In an attempt to reduce the salinity of the croplands, the soil was flushed with fresh water at least four times. This flushing also removed the nutrients that are required for crop productivity. To compensate for this loss of nutrients, local farmers began applying excessive pesticides and fertilizers. Pesticide use has increased to at least twenty times the national average and some crops from this land began exceeding maximum limits of pesticides and nitrate by two to four times. The local groundwater also suffers from the increased salt concentration, reaching six times the maximum acceptable level established by the World Health Organization.[6]

On occasion, strong northeasterly blowing windstorms entrain contaminated fine sediment and dust from the dried Aral Sea lakebed and form huge dust storms. The salt content of this dust can reach 30–40 percent in the summer but up to 90 percent in the winter. The dust storms can be as much as 93 to 186 miles (150 to 300 km) across. This contaminated dust has been found 311 to 621 miles (500 to 1,000 km) away from the Aral Sea and as far as the Russian Arctic. As a result of exposure to this wind-borne toxic dust and contaminated groundwater, the Aral Sea region is experiencing a public health crisis. Residents of the area ingest contaminants through their drinking water and by inhaling toxic dust. In addition, contaminants are absorbed by plants and the plants and water are consumed by livestock, contaminating the local food supply. Residents are suffering from a number of health problems from this continuous exposure. In the late 1990s, the infant mortality rate in the region reached 60–110 deaths per 1,000 births, compared with 48 deaths per 1,000 births in Uzbekistan and an average of 24 deaths per 1,000 births in the whole of Russia.[7]

By 2009, the infant mortality rate was still 75 deaths per 1,000 births in the region and maternity death rate was 12 deaths per 1,000 births. The Aral Sea area residents suffer from the diseases of anemia, bronchial asthma, brucellosis, and typhoid at eight times the national average. The residents have greatly elevated incidence of kidney, liver, and eye problems, digestive disorders, infectious diseases, and tuberculosis. During the 1980s, liver cancer occurrence doubled and esophageal, lung, and stomach cancer spiked. The Aral Sea disaster is estimated to have displaced at least 100,000 people and damaged the health of at least 5 million people.

The Aral Sea disaster may be the worst case of damage to surface water supply and quality but it is certainly not the only case. Many river water supplies have also been damaged especially in arid areas. Even large rivers like the Colorado River in the southwestern United States has all of its fresh water used for human consumption, and there is still not enough. But it is not only surface water. Groundwater supplies are also significantly threatened. The quality of many groundwater aquifers has been significantly compromised all over the industrialized world. But even more alarming is the rapid overuse of groundwater and depletion of aquifers in dry and agricultural areas. Much of this groundwater was accumulated in the aquifers many years ago under very different climatic conditions, such as an ice age, which will not be replicated in the foreseeable future. Many of these aquifers are in crisis.

Ogallala Aquifer Collapse

One of the most urgent and concerning groundwater supply crises is in the High Plains Aquifer, perhaps the most important aquifer in the United States. This aquifer is primarily comprised of the Ogallala Aquifer which produces about 30 percent of all agricultural irrigation water in the United States.[8] It lies within parts of eight Midwestern and southwestern states including Colorado, Kansas, Nebraska, New Mexico, Oklahoma, South Dakota, Texas, and Wyoming. The area of the aquifer is 174,000 square miles (450,000 km^2) which includes about 20 percent of the cropland for the country in one of the most important agricultural areas, possibly in the world. The water withdrawn from this aquifer is used 94 percent for irrigation to produce about 20 percent of the corn, wheat, cotton, and cattle in the United States. The groundwater in parts of the aquifer is saturated with natural contaminants which impairs the quality and safety. The percentage of these contaminants is rapidly increasing in the remaining water. However, the real crisis is the devastating removal of groundwater from the Ogallala Aquifer which is rapidly lowering the remaining water levels. This removal is creating a catastrophic situation that is currently being investigated by the US Congress.

The geologic unit that forms the Ogallala Aquifer is the result of the plate tectonic event that formed the Rocky Mountains.[9] Large rivers drained the eastern slopes of the mountains and flowed eastward carving deep valleys and channels into the preexisting rocks. They also deposited the sediments that formed the Ogallala. Gravels and coarse sands were first deposited in the river valleys and aprons of sediment were deposited across the entire area. These were mainly deposited by braided streams across the western side of the Great Plains of the central United States. The sediments covered the area reaching a thickness of about 900 feet (274 m) in some places and forming the aquifer as it was buried and saturated with groundwater.

The Ogallala Aquifer has an immense volume capable of containing groundwater.[10] The water-bearing extent of the aquifer is as much as 525 feet (160 m) thick. This thick part of the aquifer is in the northern plains. In the southern plains, it is much thinner, ranging from 50 feet (15 m) to 200 feet (61 m) thick. The depth to the groundwater is deep in the northern plains at about 400 feet (123 m) but only 100 to 200 feet (30 to 61 m) in the southern plains. Extensive pumping of water out of the Ogallala began after the Dust Bowl of the 1930s. By 1949, about 10 percent of the land over the aquifer area was irrigated for agriculture. By 1980, more than 65 percent of the aquifer area was irrigated and it has continued to increase annually. Since the 1960s, pumping of groundwater out of the aquifer has exceeded the amount of recharge. As a result of this excessive usage, the elevation of the top of the groundwater level (water table) in the Ogallala Aquifer has dropped by an average of 100 feet (30.8 m) but locally it dropped as much as 175 feet (54 m). The water table in the aquifer subsides about 1.4 feet (0.4 m) to 1.7 feet (0.5 m) every year. It is estimated that there has been a 60 percent subsidence of the water table in the Ogallala to date and it is estimated to reach 70 percent by 2060 at the current rate of removal. By natural recharge processes, it would take at least 6,000 years to replace the water that has been removed to date.

The rapid drop in the water level in the aquifer has also impacted local surface water. As a result of the loss of groundwater, the Arkansas River switched from being fed water from the aquifer through springs to losing much of its volume to the aquifer through infiltration of water through the riverbed. As a result, between 1995 and 2000, the Arkansas River lost most of its water to recharging the aquifer and it even dried up completely on occasion. The Canadian and South Platte Rivers still receive water from the Ogallala Aquifer but it is less every year. At the rate of loss of groundwater in the aquifer, these two rivers will also begin losing water to the aquifer in the near future. Even ponds, canals, and irrigation water are now feeding the aquifer in most areas as a result of this subsidence.

Rapid evaporation of the water above the Ogallala Aquifer area concentrates salts and other contaminants in the remaining water. There is also excessive usage of pesticides and fertilizers accompanying the heavy groundwater usage all designed to improve crop output. The aquifer covers such a large area that these ever-increasing contaminants permeate the soil and further contaminate the groundwater. As the surface water quality degrades from the contaminants and the infiltration rate of it increases as the water table subsides, the quality of the water in the aquifer is quickly degrading. To address this dire situation, the US Congress is considering the regulation of water usage from the Ogallala Aquifer.[11]

What can I do to help?

Almost everyone in the industrialized world can help with this problem in many ways. Certainly, direct conservation of water is the most obvious solution. Taking extended showers or showering too frequently wastes fresh water. Washing small loads of clothes or dishes uses more water than necessary. Frequent and/or extended watering of lawns or plantings wastes a lot of water especially during water shortages. Frequent washing of cars, allowing the faucets to run for no reason, frequent cleaning patios with water can all be reined in. Recreational uses of water in swimming pools, sprinklers, and other backyard devices can also waste a lot of water if used frequently. There are other less obvious ways to save water. Agriculture uses a lot of fresh water. Therefore, wasting food also wastes water. The remaining fresh water is also being degraded by releasing chemicals to the environment. Overuse of salt and other deicing compounds winds up contaminating surface and ground-water. Pesticides, fertilizers, and other agricultural chemicals for lawns and plantings also damage groundwater quality. Clearly, dumping any fuels, solvents, or other chemicals or allowing them to leak from containers also badly degrades the surface and groundwater quality. As washing machines and dishwashers wear out, they can be replaced with high-efficiency, low-water-usage units.

LOSS OF POLLINATORS

There is a quote attributed to Albert Einstein: "If the bee disappeared off the surface of the globe, then man would only have four years of life left." But it is undocumented and unproven. However, the precipitous decline in pollinator populations is most definitely a crisis that could have a serious impact on human survival.[12] It is estimated that about one third of global food production requires pollination.[13] At least ninety plant crop types in the United States require pollination for reproduction. Pollination is carried out by bees and butterflies and some types of moths, bats, and birds as well as some other types of insects. In a previous chapter, the plight of the monarch butterfly was described and most other butterflies are in equal peril. Unlike climate change, which is hard to detect in many areas, anyone living in a suburban area has noticed the disappearance of butterflies over the past forty to fifty years. Even worse is the disappearance of bees. American farmers rely on honeybees for most of the pollination of their crops. Bees swarmed every blooming flower in suburbia forty years ago. Now they are very rare with only a few in most patches of flowers. This is a problem that no one can deny. There is also an economic cost to the loss of pollinators.[14] It has been esti-mated that wild and farmed bees contribute about $15 billion in crop yields per year. This includes the more than 30 percent of the food that humans consume. The crops include all nuts and fruits including those referred to as vegetables like tomatoes. Imagine the famine if one third of our foods were no longer available. The decline in farmed bees can be tracked and the losses are documented but the loss of wild bees cannot and may be extreme.[15] From 1947 to 2008, honey-producing bee colonies decreased from almost 6 million

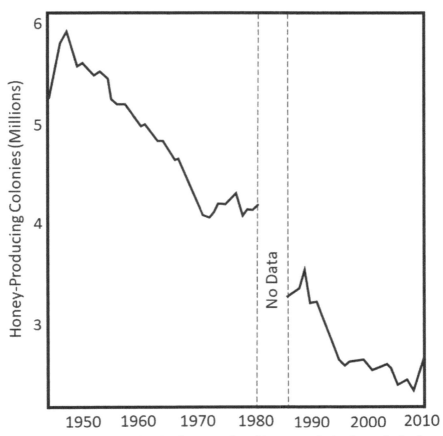

Figure 11.2. Graph showing the decrease of total honey-producing bee colonies from 1945 to 2014 in the United States using data from the US Department of Agriculture, National Agricultural Statistical Service (NASS).
Source: Adapted from Smith et al. (2014).

to about 2.3 million in the United States and they have continued to decline since then, sometimes precipitously.

Pollinator populations worldwide are in serious decline through many factors such as invasive pests like mites, viral and fungal diseases, pesticides and chemicals, loss of habitat, and loss of genetic diversity, among others.[16] Many butterfly, moth, and native bee species are extinct or nearly so in many areas.[17] Pesticides may only be one of the list of dangers but exposure to them in nonlethal doses also weakens the species in general allowing the pests and diseases to be more deadly. These pesticides are a double-edged sword. On one hand, they strongly contributed to the success of humans in the nineteenth and twentieth centuries. The infamous pesticide DDT eradicated malaria and typhus from North America and Europe and controlled the insects carrying these and other diseases worldwide. It saved hundreds of millions of human lives. Pesticides also prevent insects, plants, fungus, and animals from consuming and destroying the food supplies meant for the human population. The success in feeding the ever-increasing numbers of humans over the past two centuries greatly benefited from pesticides.

On the other hand, pesticides are very damaging to the natural environment. The factors that dictate this damage are toxicity, persistence, and mobility.[18] Toxicity is simply how efficient the chemical in the pesticide is at killing or damaging the pest. The other two factors depend heavily on the conditions of the environment into which they are released. When pesticides are applied, they can evaporate into the air or enter the natural soil and/or surface/groundwater systems. They can photodegrade to less dangerous chemicals from exposure to sunlight, settle to the soil or be washed into the water systems through precipitation and runoff from fields. Pesticides can travel all around the planet, even to remote areas like deep jungles and polar ice caps because of their persistence. Persistence is a property of the pesticide to resist chemical breakdown through reactions with air, soil, water, or in sunlight to produce less toxic by-products or removal by microbes that breakdown the pesticides. Highly persistent pesticides are typically toxic to microbes. The persistence of a pesticide is dictated by how long it takes for half of the mass of the chemical to break down to less toxic substances in the natural environment. Low-persistence pesticides have half-lives of thirty days or less, moderate persistence is thirty to a hundred days and high persistence pesticides have half-lives of more than 100 days. Pesticides with higher persistence commonly spread farther around the world than lower-persistence pesticides.

The mobility of the pesticide also dictates its impact on the natural environment.[19] Certain pesticides can adhere to soil particles which may fix them in place but if the soil is eroded by weathering, they may also be spread into surface water. Pesticides that do not adhere to soil or other particles are considered mobile. They can leach into surface and groundwater and remain active until they break down. There they may travel great distances, impacting both human health and the natural environment.

Colony Collapse Disorder

Pesticides kill or weaken multiple species of plants, insects, and other organisms in addition to the target insects. The problem with reducing target populations is that the populations of their predators are also reduced because their food supply is restricted. If the predators are food for other predators or for humans, they can eventually result in human food shortages. The danger is even greater to beneficial species like bees and other pollinators.

Wild and cultured bee populations have been declining for much of the past century. However, North American bees entered a crisis situation in 2004, when they began suffering from a new threat, colony collapse disorder (CCD).[20] Entire bee colonies began mysteriously and abruptly dying with no obvious cause. Whole colonies would disappear or if a few survivors were found, they would be malnourished and disoriented. CCD is probably the best-known case of bee decline. More recent research shows that symptoms of CCD are a very low number of adult bees in the hive but a complete absence of dead bees. Foraging and nursing bees abandon the hive. The queen bee and broods survive for a while but the absence of adult bees to take care of the babies and bring food causes them to all die. Eventually, the hive collapses. The cause of CCD is still unclear, but parasites and disease are most likely. However, it is widely believed that pesticides have weakened the bee populations through prolonged sublethal dosage making them susceptible to diseases and parasites. As a result, the pesticide imidacloprid was banned in most of the European Union as the most likely culprit. However, many other pesticides also weakened the bees among other factors.

The CCD epidemic continued to worsen with time. In 2007 alone, at least one-third of the North American bee population disappeared including about 50 percent of the Canadian bee colonies.[21, 22] CCD continued every year with various intensities. In 2015 to 2016, about 44 percent of American-managed honeybee hives perished. Similarly, in April 2020 to April 2021, American beekeepers lost about 45 percent of their honeybee hives to CCD. Managed honeybee populations are mainly remaining constant or declining slowly but native honeybees are disappearing at alarming rates. There has been some governmental response to the crisis. The National Institute of Food and Agriculture (NIFA) invested about $40 million between 2008 and 2014 to promote and protect pollinator health in the United States and continued to increase investments to about $10 million per year on pollinator research. In response to the extreme loss of pollinators, then President Obama convened a Pollinator Task Force to address the alarming loss of pollinators and to develop a national strategy to protect them in 2015.

Habitat Destruction

It is not only the sublethal doses of pesticide that are weakening pollinators so they succumb to mites and diseases. The other main factor is habitat destruction. Although there can be a few apiaries in urban areas, the natural habitat is typically so altered or replaced that only a few hardy species can survive. The only surviving insects are cockroaches, termites, fleas, ants, and possibly a few others. Normally, many pollinators should be able to survive in the ever-expanding suburban areas. However, the obsession with weed-free grass lawns and exotic ornamental plants removes most of the native plant species that pollinators depend on. The destruction of the milkweed population has been described in the plight of the monarch butterflies but the same loss is occurring with all pollinators. Anytime before about forty years ago, any blooming flower patches were so overwhelmed with pollinators that it was perilous to go near them for fear of being stung. Now in most suburbs in the United States, blooming flowers are nearly insect-free, an eerie sight. Not only have the over-applied herbicides and insecticides weakened the surviving species but the plants that they instinctually survive on are no longer there.

Recent research shows that by 2020, greater than seventy species of pollinators were endangered or threatened.[23] It found that five extraordinary or crucial native bee species were in immediate need of intervention to survive. At least 52 percent of native bee species in the United States where there was sufficient data were in serious decline and 24 percent were threatened. In Europe, the situation is no better: the number of native bee species were found to have declined by 40 percent in the United Kingdom and 60 percent in the Netherlands. Across the continent, 9.2 percent of native European bees are near extinction and 37 percent are in decline. It is estimated that almost three-quarters of agricultural crops require pollination by pollinating insects. The loss of these insects now results in inadequate pollination causing an overall 3 to 5 percent reduction in fruit, vegetable, and nut production. The lower availability of these foods causes human starvation and malnutrition resulting in about 670,000 premature deaths each year worldwide, and as pollinators continue to decline, this number will skyrocket.[24]

It is not just the butterfly and bee populations that are In such sharp decline. At least 50 percent of the bat species in the United States are endangered or in severe decline and many of them pollinate flowers as well.[25] A study in Germany showed that between 1989 and 2016, the annual bulk mass of insects caught in research traps declined by 75 percent.[26] In midsummer, when insect activity is at its peak, the decline was an astounding 82 percent. Many birds, bats, and other animals depend on these insects for food to survive. Their populations are also in freefall. Since 1980, about 122 species of amphibians worldwide have become extinct and 32 percent of those remaining are near extinction. In Central America, about 67 percent of frogs are extinct. Recent studies estimate that about 40 percent of all insects worldwide are extinct and the population reduction is 1 percent annually. Wild bird populations have decreased by 2.9 billion in North America since 1970.[27] It has been predicted that by the 2050s, the declines in insects including pollinators will range from 51 to 97 percent from original populations if nothing is done to address the situation.

What can I do to help?

Everyone can participate in addressing this dire situation at many levels. First, the production of food requires abundant use of pesticides as well as habitat destruction. All effort should be made not to waste food. This means that everyone should eat as much of the food that is purchased as possible. It also means eating only as much food as is necessary. Highly processed foods produce excessive amounts of waste so they should be replaced with less-processed foods where possible. Fruits and vegetables are also more desirable than meat because they are less detrimental to the environment in total. Second, manicured lawns and plantings also contribute to the problem. The use of herbicides and insecticides in them or around properties should be minimized or discontinued. The new trend in planting native species in a less structured landscape will help the problem immensely and should be encouraged. If pesticides must be used, they should be applied during times of day when bees and butterflies are not active to minimize their impact. However, it is far more preferable for residents to apply insect repellant to themselves or use other repellants than mass application of pesticides to kill all of the insects in an area.

EUTROPHICATION OF SURFACE WATER

The third major environmental challenge currently facing humans is eutrophication of surface waters. Excess application of chemical fertilizers in agriculture and landscaping is later removed during precipitation as runoff and enters the surface water system. The increased nutrients cause algal blooms that especially cover ponds and lakes.[28] When the algae dies, it is consumed by oxygen-consuming bacteria which progressively removes the oxygen from the water making it hypoxic. The decreased oxygen content of the water slowly chokes the aquatic life until it dies. If eutrophication occurs in streams, the mobile species will swim away to safety, if possible, while immobile organisms will perish. In ponds and lakes, none of the organisms can escape to safety. Instead, all organisms progressively perish depending upon their oxygen tolerance. Eutrophication is responsible for many of the reported fish kills in the natural environment as a result.

There are other impacts and causes of this process as well. Eutrophication can also result from excess sewage from direct dumping or from leakage from septic systems or public sewers into surface water. It can also be caused by runoff from animal feedlots and overflow of animal waste lagoons into surface water. This nutrient-rich wastewater promotes algal blooms as well. As the algae decays, it also produces high amounts of carbon dioxide which can acidify the water, lowering the pH to the point that it also damages aquatic organisms even if they survive the decreased oxygen content. In many instances, the algal blooms are blue-green algae which is really cyanobacteria. These organisms can produce toxins that are dangerous to the aquatic organisms as well as the terrestrial organisms that depend on the surface water for life or recreation. In addition, the algal blooms are on the water surface and prevent sunlight from penetrating the water. Deeper water organisms that depend on this sunlight also suffer and may die.

Dead Zones

Virtually all surface water in industrialized countries is impacted by eutrophication to some degree. A significant number of surface water bodies are basically dead as a result. However, the impacted water does not provide much of the human food supply. Therefore, minimal efforts have been made to control freshwater eutrophication. For many years experts believed that the oceans were too voluminous to be impacted by eutrophication and paid no attention. However, a large area of eutrophication developed in the saltwater Black Sea, in the USSR, in the 1960s ostensibly from natural causes. However, these saltwater eutrophication zones began occurring through runoff from over-fertilized agricultural land. These eutrophication zones were named "dead zones" because the fish and invertebrate life that usually occupied them either died or fled the area, making them devoid of appreciable life.[29] Only species that could withstand low oxygen contents remained. These dead zones began popping up with regularity all around the world through the 1980s, 1990s, and even worse in the twenty-first century. New dead zones still continue to appear today. Most occur where rivers and streams empty into large lakes or seas and oceans that have restricted circulation. Agricultural runoff washes into the rivers and streams which carry it to the ocean and it accumulates at the mouth causing eutrophication and a dead zone.[30] There are currently at least 415 dead zones identified worldwide and most of them are from over-use of agricultural fertilizer. Many of them occur in the most productive of marine areas in the world and are drastically reducing the human marine food supply.[31] If they are not controlled, they will add to food shortages and potential famines.

All the way into the 1990s, the oil production platforms in the shallow waters of the Gulf of Mexico off Louisiana and Texas used to form artificial reefs for fish. Fishermen could barely get the bait into the water before they hooked large, delicious game fish like red snapper, grouper, and pompano. It was not uncommon for small boats to bring in a catch of 100 fish or more. Now the waters are dead with not a single swimming fish. This is because they now sit in the Gulf of Mexico dead zone.[32] This dead zone appears around the mouth of the Mississippi River in February, expands through the summer when it peaks and then dissipates in the fall. In 2017, this hypoxic area grew to about 8,776 square miles (22,730 km^2) which is a little larger than the state of New Jersey but normally it is a bit smaller.

The Gulf of Mexico dead zone is produced by the influx of nitrate and some phosphate from fertilizer-rich Mississippi River water into the Gulf waters. The river carries water from thirty-one states that drain at least 41 percent of the continental United States. More than 50 percent of agricultural lands drain into the Mississippi River either directly or through its tributaries. The amount of fertilizer applied to these American agricultural fields has greatly increased since the 1950s.[33, 34] The amount of nitrate from fertilizer carried to the Gulf of Mexico tripled between 1960 and 1997 and the amount of phosphate from fertilizer doubled. By the late 1990s, 30 percent of the nitrates were from agricultural fertilizers, 30 percent was from natural soil decomposition, and 40 percent was from sewage treatment plants, animal feedlots and waste lagoons, and fallout from air pollution.

The Gulf of Mexico dead zone was first identified in the late 1970s but it was not accurately mapped until the 1990s. The zone changes size and shape and shifts its location from year to year. The zone area has been mapped accurately for the past twenty-three years. The shape, size, and location of the zone is a function of several factors such as the amount of fertilizer used in agricultural fields in the river's watershed, the amount of precipitation and resulting runoff of fertilizer-rich water in the Mississippi drainage basin, the water temperature in the Gulf of Mexico at the mouth of the river to encourage the growth of algae as well as the number of severe weather events in the Gulf which mix and dilute the fertilizer into the seawater.[35] The largest dead zones result from heavy fertilizer usage and precipitation along the river and calm, warm water in the Gulf of Mexico. In 2022, the droughts in the Midwestern United States and record low level of the Mississippi River resulted in a dead zone that was only 3,275 square miles (8,482 km^2), which was one of the smallest in recent years.

The Gulf of Mexico dead zone is a good example because there is a single, large input source but it is by far not the worst situation. The Baltic Sea lies between Scandinavia and primarily the Baltic States of Latvia, Lithuania, and Estonia.[36] It is a narrow waterway with poor circulation. Agricultural runoff has produced dead zones in the Baltic Sea since at least 1906. However, despite the cold temperatures, these dead zones have proliferated and deepened to the point that currently seven of the ten worst dead zones in the world are in the Baltic Sea.[37] The area of these dead zones reaches an astounding 27,027 square miles (70,000 km^2) at maximum, which is bigger than the state of West Virginia. Even worse, there are experts who believe that it is too late to be remediated. There are numerous other large dead zones, like the Chesapeake Bay in Maryland and Virginia, which are known for their marine productivity and are now struggling to produce catches. As these dead zones expand and deepen, they will slowly restrict and remove seafood as a human food source. At present, marine harvests produce about 17 percent of meat consumed by humans but locally, it can be 100 percent for specific communities.

What can I do to help?

Everyone can also help to reduce the number, intensities, and sizes of dead zones. Just as with the protection of pollinators, minimizing food waste reduces the need to grow more food and apply fertilizer to agricultural fields. This also follows that consuming more food than necessary and eating a diet of overprocessed foods instead of natural state foods also contributes to dead zones. Reducing or eliminating the use of fertilizers on lawns and in landscaping eliminates the amount of fertilizer in runoff. Making sure septic systems are operating properly and serving on local boards to ensure that municipal sewers are operating properly can also help. If cover crops are planted on fields in the offseasons, they slow or prevent the fertilizers from being washed off the fields by runoff which reduces the amount in rivers. This would only be possible by farmers but field experiments with this technique show sharp reductions in fertilizer runoff. Otherwise, coastal air pollution fallout can also contribute to dead zones as well.

THE TAKEAWAY

The climate crisis has attracted the attention of the world. This interest of the public in solving a major environmental problem is a welcome development. It would be a natural segue to encourage more interest in these other very serious environmental problems of fresh water supplies, loss of pollinators, and proliferation of surface water eutrophication and especially dead zones. Even though these issues appear to be minor in comparison to climate change, they are certainly comparable and some may be more important. It would be a great development to use the public interest in the climate crisis to generate or revive public interest in these other environmental crises and even additional issues not included in this analysis.

Chapter 12

If We Ignore It, Will It Go Away?

There is still a significant number of people that either do not believe that there is climate change or that the current climate change is just a natural variation and there is nothing we can do about it. If this is the case, nothing needs to be done to combat climate change and humans can proceed burning fossil fuels and harvesting forests as they have been without concern. The question is what is the worst that can happen if humans are the cause of the climate change and continue to follow this course of action? Humans have never produced enough carbon dioxide or any other greenhouse gas in the past to compare with. Therefore, there are no historical examples to gauge the potential outcome of the crisis. However, there is a natural prehistorical example that might be used as a warning. It is the great Permian extinction.

GREAT PERMIAN EXTINCTION

The great Permian extinction was the greatest mass extinction event we know of in the history of the earth and it appears to have been the result of natural climate change that cascaded into a global catastrophe. It collapsed the bio-geochemical systems of the oceans and atmosphere ultimately leading to the extinction of about 96 percent of the marine organisms and nearly 75 percent of the terrestrial organisms.[1] It took a specific set of global conditions to cause this catastrophe but the ultimate cause was erupting volcanoes that overwhelmed the atmosphere with unwanted chemicals.

The Permian is a geological period that occurred at the end of the Paleozoic Era which is the oldest of the three original geological eras.[2] The geological periods are defined by distinct forms of dominant life. The Permian had many land-dwelling reptiles and amphibians as well as rich marine life with both fish and large and advanced invertebrates. The Permian period ended about 245–250 million years ago to a world that looked much different than today. The Paleozoic was marked by the coalescing of the continents to form a single supercontinent called Pangea and one huge ocean called Panthalassa by the end of the Permian. Circulation in this huge ocean was sluggish and much weaker than it is in the many oceans of today. As the period came to an end, two major volcanic provinces developed and released huge amounts of chemicals into the atmosphere. The first event was in South China, where the Emeishan flood basalt province erupted. The lava flows from this event cover more than 96,525 square miles (250,000 km²). However, these deposits were miniscule compared to the second event called the Siberian flood basalt province or "Siberian Traps." These volcanoes began just as the South China event concluded and filled most of the huge West Siberian Basin. The volcanic lavas and pyroclastic flows covered an amazing 1.5 million square miles (3.9 million km²) which is nearly fifteen times the area of the United Kingdom. The volcanic deposits are a phenomenal 2.2 miles (3.5 km) thick. This means that the volume of magma needed to produce this massive sequence was between 0.3 and 0.6 million cubic miles (1.2–2.5 million km³).

The volcanism that produced the Siberian traps released enough CO_2 to cause a major global warming event on their own.[3] Just a single 96-cubic-mile (400 km³) lava flow of basalt can release nearly 7 gigatons (6.4 gigaMT) of CO_2 over a period of about a decade. At this rate, the volcanism that formed the Siberian traps would have released about 11,000 gigatons (10,000 gigaMT) of CO_2. This amount of CO_2 would have raised atmospheric contents by an estimated 3,000 to 7,000 parts per million (ppm) if there was no removal by the oceans or plants. There was 417 ppm in the earth's atmosphere in 2022. However, the Siberian Trap volcanism and release of CO_2 took about 75,000 to 165,000 years to complete which is geologically quick but very slow in human terms. The Siberian volcanic activity alone is calculated to have doubled the atmospheric CO_2 which would have led to a global temperature increase of between 2.7 and 8.1°F (1.5–4.5°C).[4] This does not take into account the effect of the CO_2 released by the volcanism in South China. The increased ocean temperature as the result of global warming is further speculated to have caused the release of methane gas that was sequestered in deposits on the ocean floor. Methane is about twenty-three times as strong a greenhouse gas as CO_2. This would have further increased the temperatures of the atmosphere and ocean. The estimated global temperature increase from the combined events of the late Permian period is between 9 and 12°F (5–6°C).[5]

In addition to the CO_2, the massive volcanism released large amounts of toxic chemicals as gases including chlorine, sulfur, and fluorine, which reacted with water and produced acid precipitation. This chemical change further stressed the already suffering earth's biotic systems. The ash and aerosol chemicals produced a persistent volcanic haze that reduced incoming solar radiation which reduced photosynthesis and even further stressed these systems. Much smaller historical eruptions like Laki in Iceland in 1783 caused reduction of sunlight and acid precipitation, leading to famine and shifted weather patterns that led to an estimated 6 million human deaths and possibly the French Revolution.[6] In the Permian, the change in atmospheric chemistry reduced oxygen solubility in seawater and with the increased temperature, caused the release of methane hydrates from the seafloor. All of these factors significantly reduced the efficiency of global carbon sinks allowing climate change to become a runaway situation.[7] Instead of being swept away and diluted into ocean water, the sluggish ocean circulation allowed these chemically unhealthy areas to build up. Early algal blooms were consumed by oxygen-reducing bacteria producing large areas of ocean hypoxia or "dead zones." These radical chemical changes first killed off the immobile marine life but later, as the dead zones expanded, all marine life was decimated. It appears to have taken just 10,000–30,000 years for the 96 percent of marine life to be wiped out. This indicates a complete collapse of the Earth's biologic systems and failure of the regulatory systems to stop it. This is a terrifying event.

The collapse of the Permian terrestrial biologic systems is less certain but shows that there was a strong interdependence of the marine and terrestrial systems during this catastrophe.[8] The reduced sunlight and acid precipitation from the erupting Siberian volcanoes had a devastating effect on terrestrial plants. In the geologic periods prior to the Permian, terrestrial plants had evolved explosively to produce the greatest dominance of plants in Earth's history. The lack of food from the oceans as life there went extinct caused the only mass extinction of insects in earth's history other than the current extinction. The vegetation and insects were the primary diet for larger reptiles and fish of the time. Without food, many of those species also went extinct. The larger herbivores were especially impacted because they had become dependent on an overabundance of vegetation.

COMPARISON TO THE CURRENT SITUATION

Fortunately, modern humans have been spared such devastating volcanic events. Modern human civilization is considered to have begun about 10,000 years ago, soon after the end of the last ice age. This civilization is characterized by the recording of historical events, the advancement of science and technology, organized religion, and complex political systems. Over this period, there have been only a few volcanic eruptions that altered the climate and only for a short period.[9] In addition to Laki in 1783, an unknown eruption around 535 CE and the 1815 Tambora eruption altered the climate. These events caused havoc to civilization and caused death, destruction, and shifts in society. Tambora released 38 cubic miles (160 km^3) of volcanic debris into the atmosphere that caused the year without a summer in the northern hemisphere and killed millions of people. In comparison, the Yellowstone supervolcano eruption about 630,000 years ago shot around 244 cubic miles (1,000 km^3) of ash and debris into the atmosphere. This volcanic eruption was more than six times as massive as Tambora and caused a devastating climate impact. It is possible that a volcanic eruption of this size could cause modern society to collapse.

The rate at which the climate warmed during the Permian event is estimated at 9 to 11°F (5 to 6°C) over 40,000 or 75,000 years at fastest. If the average of the changes is 10°F (5.5°C) over the average of 62,500 years, then it means that it took 11,360 years for the surface temperature to increase by 2°F (1°C.)[10] However, it is unclear at what point in this temperature increase that the extinctions began or how fast temperature changes were required to cause extinctions. Recent studies on extinctions through time suggest that temperature increases on the order of >9.5 to >18°F (>5.2 to >10°C) per million years are required for mass extinctions.[11] This means that a global temperature increase of 2°F (1°C) would take between 100,000 and 192,000 years during an extinction event.

The real cause for concern in the climate crisis is the rapidity of the current temperature increase.[12] Most yearly global temperature calculations show that the earth's climate has warmed by about 2.5°F (1.3°C) since about 1910. However, the climate heating greatly accelerated beginning in about 1980. The temperature has actually increased by about 1.8°F (1°C) over the past forty-two years and it is heating ever quicker every year. Considering that the temperature increased by 1.8°F (1°C) in 11,360 years during the worst extinction event in history, forty-two years is absolutely frightening. We have no idea what the results of this speed of heating will have on the biologic systems of the earth but it cannot be good. Even if the speed is not a factor, it will not take long to reach the absolute temperature changes of climate change-induced extinction events.

Based on the many contributing factors, models have been developed to predict temperature increases to the year 2100.[13] Although models are notoriously unreliable, they can still give a general idea of the future trends.[14] The US Global Change Research Program (USGCRP) projects that with the current rate of population increase and destruction of forests and with no action taken to reduce CO_2 emissions, by 2100, the atmospheric CO_2 levels will be at about 936 ppm and the temperature will increase by an additional 7°F (3.9°C).[15] With concerted efforts to reduce CO_2 emissions, by 2100, the levels will increase to about 550 ppm and temperatures will increase by more than 2°F (1.1°C). Even if we stop using fossil fuels today, the CO_2 levels will increase slightly by 2100. The highest CO_2 levels in the past 8,000 years was

300 ppm so the present and projected levels are cause for concern. The rates of temperature increase are also much faster than those during the Permian. It is for these reasons that it is imperative that all actions are taken to control CO_2, not just those that are convenient or designed to make money.

THE TAKEAWAY

Even if there are variations in climate change and the current trend is just a quickly heating period, the rate of temperature increase is fast beyond the point of concern. Using limited methods to address this situation is inappropriate especially when there are ways to resolve it. Just like with the other environmental crises that humans faced and overcame, people need to demand that all of these other available tools be implemented despite

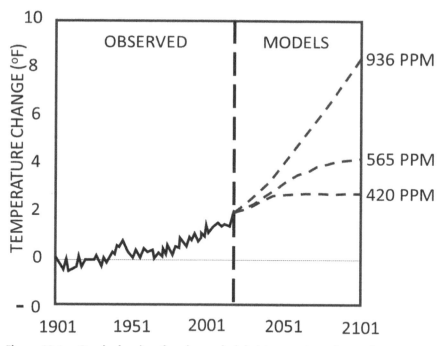

Figure 12.1. Graph showing the observed global temperature change from 1900 to present with the dotted line showing the average temperature from 1900 to 1980. The models extend to 2100 and show the upper curve for continued accelerating CO_2 emissions with minimal reduction efforts ending with 936 ppm CO_2, the middle curve of concerted efforts to reduce CO_2 emissions and ending with 550 ppm CO_2 and the lower curve with an immediate halt in CO_2 emissions ending with 420 ppm CO_2.

Source: Modified after U.S. Global Change Research Program (USGCRP), (2017).

the wishes of industry, politicians, or the wealthy. There are many ways to address climate change that are appropriate to the particular area. Further, the tools to fight climate change must take the energy needs and the time they are needed into account rather than assuming that some new technology will be developed to address a problem when it isn't really needed. In other words, the battle with the climate crisis can be won if it is fought in a reasonable manner just like the previous environmental battles. Then the other environmental problems like the freshwater shortage, loss of pollinators, and overuse of chemical fertilizers can also be addressed.

Notes

CHAPTER 1

1. Roser, Max and Hannah Ritchie, 2023. How has world population growth changed over time? *Our World in Data*.

2. Revelle, Roger, and Hans E. Suess. 1957. "Carbon dioxide exchange between atmosphere and ocean and the question of an increase of atmospheric CO2 during the past decades," *Tellus* 9 (1):18–27.

3. Gore, Al. 2006. *An Inconvenient Truth: The Planetary Emergency of Global Warming and What We Can Do About It*, Rodale.

4. EPA. "Basics of Climate Change," accessed October 22. https://www.epa.gov/climatechange-science/basics-climate-change.

5. NOAA. 2023. "The Atmosphere," last modified July 28, 2023, accessed October 22. https://www.noaa.gov/jetstream/atmosphere.

6. NASA. "Earth's Moon." accessed October 22. https://moon.nasa.gov/inside-and-out/dynamic-moon/weather-on-the-moon/.

7. Squyres, S. W. "Venus." Encyclopedia Britannica, last modified October 19, 2023, accessed October 22, 2023. https://www.britannica.com/place/Venus-planet.

8. NASA. 2022. "Global Temperature." accessed October 23. https://climate.nasa.gov/vital-signs/global-temperature/.

9. Lindsey, Rebecca. 2023. "Climate Change: Atmospheric Carbon Dioxide." Climate.gov, accessed October 23. https://www.climate.gov/news-features/understanding-climate/climate-change-atmospheric-carbon-dioxide.

10. Raynaud, Dominique, Jai Chowdhry Beeman, Jérome Chappellaz, F. Parrenin, and Jinhwa Shin. 2020. The long-term ice core record of CO2 and other greenhouse gases.

11. Revelle, Roger, and Hans E. Suess. 1957. "Carbon dioxide exchange between atmosphere and ocean and the question of an increase of atmospheric CO2 during the past decades." *Tellus* 9 (1):18–27.

12. Mann, Michael E. 2002. "Little ice age." *Encyclopedia of global environmental change* 1 (504):e509.

13. Edgerton-Tarpley, Kathryn. 2016. North China famine, 1876–79.

14 . Gates, Alexander. 2023. *Polluted Earth: The Science of the Earth's Environment*: John Wiley & Sons.

15. Benton, Michael J, and Richard J Twitchett. 2003. "How to kill (almost) all life: the end-Permian extinction event." *Trends in Ecology & Evolution* 18 (7):358–65.

16. Lindsey, Rebecca. 2022. "Climate Change: Global Sea Level." Climate.gov, accessed October 23. https://www.climate.gov/news-features/understanding-climate/climate-change-global-sea-level.

17. Oda, Takahiro, Jun'ya Takakura, Longlong Tang, Toshichika Iizumi, Norihiro Itsubo, Haruka Ohashi, Masashi Kiguchi, Naoko Kumano, Kiyoshi Takahashi, and Masahiro Tanoue. 2023. "Total economic costs of climate change at different discount rates for market and non-market values." *Environmental Research Letters* 18 (8):084026.

18. Lindsey, Rebecca and Dahlman, LuAnn 2023. "Climate Change: Ocean Heat Content." climate.gov, accessed October 23. https://preview.climate.gov/news-features/understanding-climate/climate-change-ocean-heat-content.

19. Gates, Alexander. 2023. *Polluted Earth: The Science of the Earth's Environment*: John Wiley & Sons.

20. Erdman, Jonathan and Wesner Childs, January 2023. "Hurricane Ian: Lessons Learned One Year Later." The Weather Channel, accessed October 23. https://weather.com/safety/hurricane/news/2023-09-21-florida-hurricane-ian-lessons-learned.

21. EPA. 2021. "Climate Change Indicators: Tropical Cyclone Activity." Last Modified July 21, 2023, accessed October 24. https://www.epa.gov/climate-indicators/climate-change-indicators-tropical-cyclone-activity.

22. Buchholtz, K. 2020. "2010s see Record Number of Storms in the Philippines." Statista.com, accessed October 24. https://www.statista.com/chart/23638/storms-storm-deaths-in-the-philippines/.

23. NOAA. 2011. "Extended Multivariate ENSO Index (MEI.ext)." accessed October 24. https://psl.noaa.gov/enso/mei.ext/.

24. EPA. 2022. "Climate Change Indicators: Wildfires." EPA, accessed October 24. https://www.epa.gov/climate-indicators/climate-change-indicators-wildfires.

25. Patel, K. 2018. "Six trends to know about fire season in the western U.S." NASA Global Climate Change, accessed October 25. https://climate.nasa.gov/explore/ask-nasa-climate/2830/six-trends-to-know-about-fire-season-in-the-western-us/.

26. Gates, Alexander. 2023. *Polluted Earth: The Science of the Earth's Environment*: John Wiley & Sons.

27. NOAA. 2022. "Warm, dry October intensifies U.S. drought," accessed October 25. https://www.noaa.gov/news/warm-dry-october-intensifies-us-drought.

28. Copernicus. 2022. "River Discharge." European State of the Climate, accessed October 25. https://climate.copernicus.eu/esotc/2022/river-discharge.

CHAPTER 2

1. US Fish & Wildlife Service. "Rachel Carson (1907–1964) Author of the Modern Environmental Movement." Accessed October 26. https://www.fws.gov/staff-profile/rachel-carson-1907-1964-author-modern-environmental-movement.

2. Carson, Rachel. 2009. *Silent Spring*. 1962. Los Angeles, CA: Getty Publications.

3. Gates, Alexander E., and Robert P. Blauvelt. 2011. *Encyclopedia of Pollution*, Facts on File.

4. Gates, Alexander. 2023. *Polluted Earth: The Science of the Earth's Environment*, John Wiley & Sons.

5. Agency for Toxic Substances and Disease Registry. 2022. "Toxicological Profile for DDT, DDE, and DDD." US Centers for Disease Control, accessed October 26. https://www.atsdr.cdc.gov/toxprofiles/tp35.pdf.

6. Turusov, Vladimir, Valery Rakitsky, and Lorenzo Tomatis. 2002. "Dichloro-diphenyltrichloroethane (DDT): ubiquity, persistence, and risks." *Environmental Health Perspectives* 110 (2):125–28.

7. Gates, Alexander E., and Robert P. Blauvelt. 2011. *Encyclopedia of Pollution*, Facts on File.

8. Graham, F. Jr. 1978. "Rachel Carson." EPA, accessed October 26. https://www.epa.gov/archive/epa/aboutepa/rachel-carson.html.

9. Carson, Rachel. 2009. *Silent Spring*. 1962. Los Angeles, CA: Getty Publications.

10. Carson, Rachel. 2009. *Silent Spring*. 1962. Los Angeles, CA: Getty Publications.

11. Gates, Alexander. 2023. *Polluted Earth: The Science of the Earth's Environment*, John Wiley & Sons.

12. Agency for Toxic Substances and Disease Registry. 2022. "Toxicological Profile for DDT, DDE, and DDD." US Centers for Disease Control, accessed October 26. https://www.atsdr.cdc.gov/toxprofiles/tp35.pdf.

13. Gates, Alexander. 2023. *Polluted Earth: The Science of the Earth's Environment*, John Wiley & Sons.

14. Smith, Daniel. 1999. "Worldwide trends in DDT levels in human breast milk." *International Journal of Epidemiology* 28 (2):179–88.

15. Calabrese, Edward J. 1982. "Human breast milk contamination in the United States and Canada by chlorinated hydrocarbon insecticides and industrial pollutants: current status." *Journal of the American College of Toxicology* 1 (3):91–98.

16. Savage, E. P., T. J. Keefe, J. D. Tessari, H. W. Wheeler, F. M. Applehans, E. A. Goes, and S. A. Ford. 1981. "National study of chlorinated hydrocarbon insecticide residues in human milk, USA: I. Geographic distribution of dieldrin, heptachlor, heptachlor epoxide, chlordane, oxychlordane, and mirex." *American Journal of Epidemiology* 113 (4):413–22.

17. Dillon, J.-C., G. B. Martin, and H. T. O'Brien. 1981. "Pesticide residues in human milk." *Food and Cosmetics Toxicology* 19:437–42.

18. Rogan, W. J., Bagniewska, A., and Damstra, T. 1980. "Pollutants in breast milk." *New England Journal of Medicine* 302:1450–53.

19. Agency for Toxic Substances and Disease Registry. 2022. "Toxicological Profile for DDT, DDE, and DDD." US Centers for Disease Control, accessed October 26. https://www.atsdr.cdc.gov/toxprofiles/tp35.pdf.

20. Berry-Caban, Cristobal S. 2011. "DDT and silent spring: fifty years after." *Journal of Military Veteran Health* 19:19–24.

21. Rogan, W. J., Bagniewska, A., and Damstra, T. 1980. "Pollutants in breast milk." *New England Journal of Medicine* 302:1450–53.

22. Carson, Rachel. 2009. *Silent Spring*. 1962. Los Angeles, CA: Getty Publications.

23. U.S Fish & Wildlife Service. 2021. "Fact sheet: Bald Eagle Haliaeetus leucocephalus." Accessed October 26. https://www.fws.gov/sites/default/files/documents/bald-eagle-fact-sheet.pdf.

24. Gates, Alexander. 2023. *Polluted Earth: The Science of the Earth's Environment*, John Wiley & Sons.

25. Gates, Alexander E., and Robert P. Blauvelt. 2011. *Encyclopedia of Pollution*, Facts on File.

26. Gates, Alexander. 2023. *Polluted Earth: The Science of the Earth's Environment*: John Wiley & Sons.

27. U.S Fish & Wildlife Service. 2021. "Fact sheet: Bald Eagle Haliaeetus leucocephalus." Accessed October 26. https://www.fws.gov/sites/default/files/documents/bald-eagle-fact-sheet.pdf.

28. U.S Fish & Wildlife Service. 2021. "Fact sheet: Bald Eagle Haliaeetus leucocephalus." Accessed October 26. https://www.fws.gov/sites/default/files/documents/bald-eagle-fact-sheet.pdf.

29. U.S Fish & Wildlife Service. 1999. "Peregrine Falcon (Falco peregrinus)." Accessed October 26. http://npshistory.com/brochures/nwr/wildlife-fact-sheets/peregrine-falcon-1999.pdf.

30. U.S Fish & Wildlife Service. 2009. "Fact sheet: Brown Pelican Pelecanus occidentalis." Accessed October 26. https://www.fws.gov/sites/default/files/documents/brown_pelicanfactsheet09.pdf.

CHAPTER 3

1. Agency for Toxic Substances and Disease Registry (ATSDR). 2022. "ATSDR's Substance Priority List." Accessed October 26. https://www.atsdr.cdc.gov/spl/index.html.

2. Gates, Alexander E., and Robert P. Blauvelt. 2011. *Encyclopedia of Pollution*, Facts on File.

3. Gates, Alexander. 2023. *Polluted Earth: The Science of the Earth's Environment*, John Wiley & Sons.

4. Nriagu, Jerome O. 1983. *Lead and Lead Poisoning in Antiquity*. New York and Chichester: J. Wiley.

5. Needleman, Herbert L. 1999. "History of lead poisoning in the world." International conference on lead poisoning prevention and treatment, Bangalore.

6. Waldron, Harry A. 1973. "Lead poisoning in the ancient world." *Medical History* 17 (4):391–99.

7. Needleman, Herbert L. 1999. "History of lead poisoning in the world." International conference on lead poisoning prevention and treatment, Bangalore.

8. Rich, V. 1994. *The International Lead Trade*, Woodhead Publishing.

9. Needleman, Herbert L. 1999. "History of lead poisoning in the world." International conference on lead poisoning prevention and treatment, Bangalore.

10. Delile, Hugo, Janne Blichert-Toft, Jean-Philippe Goiran, Simon Keay, and Francis Albarède. 2014. "Lead in ancient Rome's city waters." *Proceedings of the National Academy of Sciences* 111 (18):6594–6599.

11. Waldron, Harry A. 1973. "Lead poisoning in the ancient world." *Medical History* 17 (4):391–99.

12. Waldron, Harry A. 1973. "Lead poisoning in the ancient world." *Medical History* 17 (4):391–99.

13. Needleman, Herbert L. 1999. "History of lead poisoning in the world." International conference on lead poisoning prevention and treatment, Bangalore.

14. Rich, V. 1994. *The International Lead Trade*, Woodhead Publishing.

15. Lewis, J. 1985. "Lead poisoning: A historical perspective." *EPA Journal.*

16. Lewis, J. 1985. "Lead poisoning: A historical perspective." *EPA Journal.*

17. Rich, V. 1994. *The International Lead Trade*, Woodhead Publishing.

18. Lewis, J. 1985. "Lead poisoning: A historical perspective." *EPA Journal.*

19. Rich, V. 1994. *The International Lead Trade*, Woodhead Publishing.

20. Gates, Alexander E., and Robert P. Blauvelt. 2011. *Encyclopedia of Pollution*, Facts on File.

21. Lewis, J. 1985. "Lead poisoning: A historical perspective." *EPA Journal.*

22. Agency for Toxic Substances and Disease Registry. 2020. "Toxicological Profile for Lead." Accessed October 26. https://www.atsdr.cdc.gov/ToxProfiles/tp13.pdf.

23. Reuben, Aaron, Maxwell Elliott, and Avshalom Caspi. 2020. "Implications of legacy lead for children's brain development." *Nature Medicine* 26 (1):23–25.

24. Gates, Alexander E., and Robert P. Blauvelt. 2011. *Encyclopedia of Pollution*, Facts on File.

25. Lewis, J. 1985. "Lead poisoning: A historical perspective." *EPA Journal.*

26. Tilton, George. 1998. "Clair Cameron Patterson." In *Biographical Memoirs*, 266–87. Washington, DC: The National Academies Press.

27. Tilton, George. 1998. "Clair Cameron Patterson." In *Biographical Memoirs*, 266–87. Washington, DC: The National Academies Press.

28. Nriagu, Jerome O. 1983. *Lead and Lead Poisoning in Antiquity.* New York and Chichester: J. Wiley.

29. Tilton, George. 1998. "Clair Cameron Patterson." In *Biographical Memoirs*, 266–87. Washington, DC: The National Academies Press.

30. Gates, Alexander E., and Robert P. Blauvelt. 2011. *Encyclopedia of Pollution*, Facts on File.

31. Nriagu, Jerome O. 1983. *Lead and Lead Poisoning in Antiquity.* New York and Chichester: J. Wiley.

32. Tilton, George. 1998. "Clair Cameron Patterson." In *Biographical Memoirs*, 266–87. Washington, DC: The National Academies Press.

33. Nriagu, Jerome O. 1983. *Lead and Lead Poisoning in Antiquity*. New York and Chichester: J. Wiley.

34. Lewis, J. 1985. "Lead poisoning: A historical perspective." *EPA Journal.*

35. Agency for Toxic Substances and Disease Registry. 2020. "Toxicological Profile for Lead." Accessed October 26. https://www.atsdr.cdc.gov/ToxProfiles/tp13.pdf.

36. Tilton, George. 1998. "Clair Cameron Patterson." In *Biographical Memoirs*, 266–87. Washington, DC: The National Academies Press.

37. Patterson, C., Shirahata, H. and Ericson J. 1987. "Lead in ancient human bones and its relevance to historical developments of social problems with lead." *Science of the Total Environment* 61:167–200. doi: 10.1016/0048-9697(87)90366-4.

38. Gates, Alexander. 2023. *Polluted Earth: The Science of the Earth's Environment*, John Wiley & Sons.

39. Ritchie, Hannah, and Max Roser. 2022. "Lead pollution." *Our World in Data.*

40. Gates, Alexander E., and Robert P. Blauvelt. 2011. *Encyclopedia of Pollution*, Facts on File.

41. Nriagu, Jerome O. 1983. *Lead and Lead Poisoning in Antiquity*. New York and Chichester: J. Wiley.

42. Gates, Alexander. 2023. *Polluted Earth: The Science of the Earth's Environment*, John Wiley & Sons.

43. Idaho Department of Environmental Quality (IDEQ). "Bunker Hill Superfund Site." Accessed October 28. https://www.deq.idaho.gov/waste-management-and-remediation/mining-in-idaho/bunker-hill-superfund-site/.

44. EPA. "Superfund Site: Bunker Hill Mining & Metallurgical Complex Smelterville, ID." Accessed October 26. https://cumulis.epa.gov/supercpad/cursites/csitinfo.cfm?id=1000195&msspp=med.

45. Idaho Department of Environmental Quality. "Bunker Hill Superfund Site." Accessed October 28. https://www.deq.idaho.gov/waste-management-and-remediation/mining-in-idaho/bunker-hill-superfund-site/.

46. EPA. "Superfund Site: Bunker Hill Mining & Metallurgical Complex Smelterville, ID." Accessed October 26. https://cumulis.epa.gov/supercpad/cursites/csitinfo.cfm?id=1000195&msspp=med.

47. EPA. "Superfund Site: Bunker Hill Mining & Metallurgical Complex Smelterville, ID." Accessed October 26. https://cumulis.epa.gov/supercpad/cursites/csitinfo.cfm?id=1000195&msspp=med.

48. Denchak, M. 2018. "Flint water crisis: Everything you need to know." Natural Resources Defense Council. https://www.nrdc.org/stories/flint-water-crisis-everything-you-need-know.

49. Nriagu, Jerome O. 1983. *Lead and Lead Poisoning in Antiquity*. New York and Chichester: J. Wiley.

50. Gates, Alexander. 2023. *Polluted Earth: The Science of the Earth's Environment*, John Wiley & Sons.

51. Denchak, M. 2018. "Flint water crisis: Everything you need to know." Natural Resources Defense Council. https://www.nrdc.org/stories/flint-water-crisis -everything-you-need-know.

52. Centers for Disease Control and Prevention. 2016. "Flint water crisis." Last Modified October 24, 2023. https://www.cdc.gov/nceh/casper/pdf-html/flint_water _crisis_pdf.html.

53. Denchak, M. 2018. "Flint water crisis: Everything you need to know." Natural Resources Defense Council NDRC. https://www.nrdc.org/stories/flint-water-crisis -everything-you-need-know.

54. EPA. "Flint drinking water response." Last Modified October 24, 2023, accessed October 26. https://www.epa.gov/flint.

55. EPA. "Flint drinking water response." Last Modified October 24, 2023, accessed October 26. https://www.epa.gov/flint.

56. Centers for Disease Control and Prevention. 2016. "Flint water crisis." Last Modified October 24, 2023. https://www.cdc.gov/nceh/casper/pdf-html/flint_water _crisis_pdf.html.

57. Nriagu, Jerome O. 1983. *Lead and Lead Poisoning in Antiquity*. New York and Chichester: J. Wiley.

CHAPTER 4

1. Gates, Alexander E., and Robert P. Blauvelt. 2011. *Encyclopedia of Pollution*, Facts on File.

2. Lippmann, Morton. 1989. "Health effects of ozone a critical review." *Japca* 39 (5):672–95.

3. EPA. "Basic Ozone Layer Science." Last Modified October 7, 2021, accessed October 28. https://www.epa.gov/ozone-layer-protection/basic-ozone-layer-science.

4. Gates, Alexander E., and Robert P. Blauvelt. 2011. *Encyclopedia of Pollution*, Facts on File.

5. EPA. "Basic Ozone Layer Science." Last Modified October 7, 2021, accessed October 28. https://www.epa.gov/ozone-layer-protection/basic-ozone-layer-science.

6. Slaper, Harry, Guus J. M. Velders, John S. Daniel, Frank R. de Gruijl, and Jan C. van der Leun. 1996. "Estimates of ozone depletion and skin cancer incidence to examine the Vienna Convention achievements." *Nature* 384 (6606):256–58.

7. Umar, Sheikh Ahmad, and Sheikh Abdullah Tasduq. 2022. "Ozone layer depletion and emerging public health concerns—an update on epidemiological perspective of the ambivalent effects of ultraviolet radiation exposure." *Frontiers in Oncology* 12:866733.

8. Gates, Alexander E., and Robert P. Blauvelt. 2011. *Encyclopedia of Pollution*, Facts on File.

9. Gates, Alexander E., and Robert P. Blauvelt. 2011. *Encyclopedia of Pollution*, Facts on File.

10. Gantz, Carroll. 2015. *Refrigeration: A History*, McFarland.

11. Tsai, W. T. 2014. Chlorofluorocarbons. In *Encyclopedia of Toxicology*, edited by Philip Wexler, Academic Press.

12. Gantz, Carroll. 2015. *Refrigeration: A History*, McFarland.

13. Gates, Alexander E., and Robert P. Blauvelt. 2011. *Encyclopedia of Pollution*, Facts on File.

14. Gates, Alexander E., and Robert P. Blauvelt. 2011. *Encyclopedia of Pollution*, Facts on File.

15. Molina, M. and Rowland, S. 1974. "Stratospheric sink for chlorofluorocarbons: chlorine atom-catalysed destruction of ozone." *Nature* 249:810–12.

16. American Chemical Society. 2017. "Chlorofluorocarbons and Ozone Depletion." Accessed October 28. http://www.acs.org/content/acs/en/education/whatischemistry/landmarks/cfcs-ozone.html

17. World Meteorological Organization. 2018. Scientific Assessment of Ozone Depletion: 2018. In *Global Ozone Research and Monitoring Project*, Geneva, Switzerland.

18. Tsai, W. T. 2014. Chlorofluorocarbons. In *Encyclopedia of Toxicology*, edited by Philip Wexler, Academic Press.

19. Williamson, Marcus. 2012. Professor Sherwood Rowland Scientist who helped establish CFCs' harmful effects. *Independent*. Accessed October 28.

20. Council, National Research. 1976. *Halocarbons: Effects on Stratospheric Ozone*, National Academy of Sciences.

21. Fitzpatrick, T. B. 1976. "Halocarbons: Environmental effects of chlorofluoromethane release." Committee on Impacts of Stratospheric Change. Assembly of Mathematical and Physical Sciences, National Research Council. Washington, DC: National Academy of Sciences, 123.

22. Farman, Joseph C., Brian G. Gardiner, and Jonathan D. Shanklin. 1985. "Large losses of total ozone in Antarctica reveal seasonal ClO x/NO x interaction." *Nature* 315 (6016):207–10.

23. World Meteorological Organization. 2018. Scientific Assessment of Ozone Depletion: 2018. In *Global Ozone Research and Monitoring Project*, Geneva, Switzerland.

24. U.S. Department of State. The Montreal Protocol on substances that deplete the ozone layer.

25. Noakes, Tim J. 1995. "CFCs, their replacements, and the ozone layer." *Journal of Aerosol Medicine* 8 (s1):S-3-S-7.

26. NOAA. 2021. "Antarctic ozone hole is 13th largest on record and expected to persist into November," accessed October 28. https://www.noaa.gov/news/antarctic-ozone-hole-is-13th-largest-on-record-and-expected-to-persist-into-november.

27. NASA. "NASA Ozone Watch." accessed October 28. https://ozonewatch.gsfc.nasa.gov/SH.html.

28. Blakemore, Erin. 2016. "The Ozone Hole Was Super Scary, So What Happened to It?" *Smithsonian Magazine*.

CHAPTER 5

1. Gates, Alexander. 2023. *Polluted Earth: The Science of the Earth's Environment*, John Wiley & Sons.

2. Gates, Alexander. 2023. *Polluted Earth: The Science of the Earth's Environment*, John Wiley & Sons.

3. Gates, Alexander E., and David Ritchie. 2006. *Encyclopedia of Earthquakes and Volcanoes*, Infobase Publishing.

4. Gates, Alexander E., and David Ritchie. 2006. *Encyclopedia of Earthquakes and Volcanoes*, Infobase Publishing.

5. Thordarson, Thorvaldur, and Stephen Self. 2003. "Atmospheric and environmental effects of the 1783–1784 Laki eruption: A review and reassessment." *Journal of Geophysical Research: Atmospheres* 108 (D1):AAC 7–1-AAC 7–29.

6. Gates, Alexander. 2023. *Polluted Earth: The Science of the Earth's Environment*, John Wiley & Sons.

7. Gates, Alexander. 2023. *Polluted Earth: The Science of the Earth's Environment*, John Wiley & Sons.

8. Gates, Alexander E., and Robert P. Blauvelt. 2011. *Encyclopedia of Pollution*, Facts on File.

9. Selby, Karen. "Mesothelioma death and mortality rate." Asbestos.com, last modified September 29, 2023, accessed October 29. https://www.asbestos.com/mesothelioma/death-rate/.

10. Gates, Alexander E., and Robert P. Blauvelt. 2011. *Encyclopedia of Pollution*, Facts on File.

11. Vallero, Daniel A. 2014. *Fundamentals of Air Pollution*, Academic press.

12. Gates, Alexander E., and Robert P. Blauvelt. 2011. *Encyclopedia of Pollution*, Facts on File.

13. Kim, Ki-Hyun, Ehsanul Kabir, and Shamin Kabir. 2015. "A review on the human health impact of airborne particulate matter." *Environment International* 74:136–43.

14. Nemery, Benoit, Peter HM Hoet, and Abderrahim Nemmar. 2001. "The Meuse Valley fog of 1930: An air pollution disaster." *The Lancet* 357 (9257):704–8.

15. Jacobs, Elizabeth T., Jefferey L. Burgess, and Mark B. Abbott. 2018. "The Donora smog revisited: 70 years after the event that inspired the Clean Air Act." *American Journal of Public Health* 108 (S2):S85–S88.

16. Brimblecombe, P. 2012. *The Big Smoke: A History of Air Pollution in London since Medieval Times*, Taylor & Francis.

17. Gates, Alexander E., and Robert P. Blauvelt. 2011. *Encyclopedia of Pollution*, Facts on File.

18. Gates, Alexander E., and Robert P. Blauvelt. 2011. *Encyclopedia of Pollution*, Facts on File.

19. Rothschild, Rachel Emma. 2019. *Poisonous Skies: Acid Rain and the Globalization of Pollution*, University of Chicago Press.

20. ATSDR. 1998. "Toxicological Profile for Sulfur Dioxide." U.S. Dept of Health and Human Services, accessed October 30. https://www.atsdr.cdc.gov/toxprofiles/tp116.pdf.

21. ATSDR. 2002. "ToxFAQs for Nitrogen Oxides." U.S. Department of Health and Human Services, accessed October 30. https://wwwn.cdc.gov/TSP/ToxFAQs/ToxFAQsDetails.aspx?faqid=396&toxid=69.

22. Katoh, T., T. Konno, I. Koyama, H. Tsurata, and H. Makino. 1990. "Acidic Precipitation in Japan." In *Acidic Precipitation: International Overview and Assessment*, edited by A. H. M. Bresser and W. Salomons, 41–105. New York: Springer.

23. Rothschild, Rachel Emma. 2019. *Poisonous Skies: Acid Rain and the Globalization of Pollution*, University of Chicago Press.

24. Vallero, Daniel A. 2014. *Fundamentals of Air Pollution*, Academic Press.

25. Hoesly, Rachel M., Steven J. Smith, Leyang Feng, Zbigniew Klimont, Greet Janssens-Maenhout, Tyler Pitkanen, Jonathan J. Seibert, Linh Vu, Robert J. Andres, and Ryan M. Bolt. 2018. "Historical (1750–2014) anthropogenic emissions of reactive gases and aerosols from the Community Emissions Data System (CEDS)." *Geoscientific Model Development* 11 (1):369–408.

26. EPA. "NAAQS Table," last modified March 15, 2023, accessed October 29. https://www.epa.gov/criteria-air-pollutants/naaqs-table.

27. Irving, P. M. 1988. "Overview of the U.S. National Acid Precipitation Assessment Program." Dordrecht.

28. Schmalensee, Richard, and Robert N Stavins. 2019. "Policy evolution under the clean air act." *Journal of Economic Perspectives* 33 (4):27–50.

29. Gates, Alexander E., and Robert P. Blauvelt. 2011. *Encyclopedia of Pollution*, Facts on File.

30. Johnsen, Reid, Jacob LaRiviere, and Hendrik Wolff. 2019. "Fracking, coal, and air quality." *Journal of the Association of Environmental and Resource Economists* 6 (5):1001–1037.

31. EPA. "Sulfur Dioxide Trends." Last modified May 23, 2023, accessed October 28. https://www.epa.gov/air-trends/sulfur-dioxide-trends.

32. Hoesly, Rachel M., Steven J. Smith, Leyang Feng, Zbigniew Klimont, Greet Janssens-Maenhout, Tyler Pitkanen, Jonathan J. Seibert, Linh Vu, Robert J. Andres, and Ryan M. Bolt. 2018. "Historical (1750–2014) anthropogenic emissions of reactive gases and aerosols from the Community Emissions Data System (CEDS)." *Geoscientific Model Development* 11 (1):369–408.

33. Gates, Alexander E., and Robert P. Blauvelt. 2011. *Encyclopedia of Pollution*, Facts on File.

34. EPA. "Nitrogen Dioxide Trends," accessed October 28. https://www.epa.gov/air-trends/nitrogen-dioxide-trends.

35. EPA. "Nitrogen Dioxide Trends," accessed October 28. https://www.epa.gov/air-trends/nitrogen-dioxide-trends.

36. Hoesly, Rachel M., Steven J. Smith, Leyang Feng, Zbigniew Klimont, Greet Janssens-Maenhout, Tyler Pitkanen, Jonathan J. Seibert, Linh Vu, Robert J. Andres, and Ryan M. Bolt. 2018. "Historical (1750–2014) anthropogenic emissions of

reactive gases and aerosols from the Community Emissions Data System (CEDS)." *Geoscientific Model Development* 11 (1):369–408.

CHAPTER 6

1. U.S. Energy Information Administration. "Oil and Petroleum Products Explained." U.S. Department of Energy, last modified June 12, 2023, accessed October 31. https://www.eia.gov/energyexplained/oil-and-petroleum-products/.

2. U.S. Energy Information Administration. "Oil and Petroleum Products Explained." U.S. Department of Energy, last modified June 12, 2023, accessed October 31. https://www.eia.gov/energyexplained/oil-and-petroleum-products/.

3. U.S. Energy Information Administration. "Oil and Petroleum Products Explained." U.S. Department of Energy, last modified June 12, 2023, accessed October 31. https://www.eia.gov/energyexplained/oil-and-petroleum-products/.

4. Solly, Ray. 2022. "The Development of Crude Oil Tankers: A Historical Miscellany." *The Development of Crude Oil Tankers*, 1–192.

5. Burger, Joanna. 1997. *Oil Spills*. New Brunswick, NJ: Rutgers University Press.

6. Wells, P. G. 2017. "The iconic Torrey Canyon oil spill of 1967—Marking its legacy." *Marine Pollution Bulletin* 115 (1/2):1–2.

7. Vaughn, A. 2017. Torrey Canyon disaster—the UK's worst-ever oil spill 50 years on. Accessed October 31, 2023.

8. Bell, Bethan, and Mario Cacciottolo. 2017. Torrey Canyon oil spill: The day the sea turned black.

9. Petrow, Richard. 1968. *In the Wake of Torrey Canyon*: D. McKay Company.

10. Vaughn, A. 2017. Torrey Canyon disaster—the UK's worst-ever oil spill 50 years on. Accessed October 31, 2023.

11. Bell, Bethan, and Mario Cacciottolo. 2017. Torrey Canyon oil spill: The day the sea turned black.

12. Burger, Joanna. 1997. *Oil Spills*. New Brunswick, NJ: Rutgers University Press.

13. Cedre. "Atlantic Empress/Aegean Captain." Cedre, Last Modified 08/02/2007, accessed October 31. https://wwz.cedre.fr/en/Resources/Spills/Spills/Atlantic-Empress-Aegean-Captain.

14. Gates, Alexander E., and Robert P. Blauvelt. 2011. *Encyclopedia of Pollution*, Facts on File.

15. Horn, Stuart A., and Captain Phillip Neal. 1981. "The Atlantic Empress sinking—a large spill without environmental disaster." International Oil Spill Conference.

16. Horn, Stuart A., and Captain Phillip Neal. 1981. "The Atlantic Empress sinking—a large spill without environmental disaster." International Oil Spill Conference.

17. Horn, Stuart A., and Captain Phillip Neal. 1981. "The Atlantic Empress sinking—a large spill without environmental disaster." International Oil Spill Conference.

18. Mambra, S. 2022. The complete story of the Exxon Valdez oil spill. *Maritime History* 2023 (October 31).

19. Gates, Alexander E., and Robert P. Blauvelt. 2011. *Encyclopedia of Pollution*, Facts on File.

20. Mambra, S. 2022. The complete story of the Exxon Valdez oil spill. *Maritime History* 2023 (October 31).

21. Commission, Alaska Oil Spill. 1990. Spill: The wreck of the Exxon Valdez, Final Report. Alaska: State of Alaska.

22. Commission, Alaska Oil Spill. 1990. Spill: The wreck of the Exxon Valdez, Final Report. Alaska: State of Alaska.

23. Commission, Alaska Oil Spill. 1990. Spill: The wreck of the Exxon Valdez, Final Report. Alaska: State of Alaska.

24. Galt, Jerry A., William J. Lehr, and Debra L. Payton. 1991. "Fate and transport of the Exxon Valdez oil spill. Part 4." *Environmental Science & Technology* 25 (2):202–9.

25. Galt, Jerry A., William J. Lehr, and Debra L. Payton. 1991. "Fate and transport of the Exxon Valdez oil spill. Part 4." *Environmental Science & Technology* 25 (2):202–9.

26. Struck, D. 2009. Twenty Years Later, Impacts of the Exxon Valdez Linger. *Yale Environment 360*.

27. Gates, Alexander E., and Robert P. Blauvelt. 2011. *Encyclopedia of Pollution*, Facts on File.

28. Morgan, J. D. 2011. "The Oil Pollution Act of 1990." *Fordham Environmental Law Review* 6 (1):1–27.

29. (ITOPF), International Tanker Owners Pollution Federation. 2023. "Oil tanker spill statistics 2022." ITOPF Ltd., accessed October 31. https://www.itopf.org/knowledge-resources/data-statistics/statistics/.

CHAPTER 7

1. Juuti, Petri S., Tapio Katko, and Heikkis Vuorinen. 2007. *Environmental History of Water*, IWA Publishing.

2. Gates, Alexander E., and Robert P. Blauvelt. 2011. *Encyclopedia of Pollution*, Facts on File.

3. Angelakis, Andreas N., Andrea G. Capodaglio, Cees W. Passchier, Mohammad Valipour, Jens Krasilnikoff, Vasileios A. Tzanakakis, Gül Sürmelihindi, Alper Baba, Rohitashw Kumar, and Benoît Haut. 2023. "Sustainability of water, sanitation, and hygiene: From prehistoric times to the present times and the future." *Water* 15 (8):1614.

4. Deming, David. 2020. "The aqueducts and water supply of Ancient Rome." *Ground Water* 58 (1):152.

5. Havlíček, Filip, and Miroslav Morcinek. 2016. "Waste and pollution in the ancient Roman Empire." *Journal of Landscape Ecology* 9 (3):33–49.

6. Joel, L. 2018. "Ancient Romans polluted their lakes just like we do today." *Eos* 99. Accessed November 4, 2023. doi:10.1029/2018EO110747.

7. Deming, David. 2020. "The aqueducts and water supply of Ancient Rome." *Ground Water* 58 (1):152.

8. Havlíček, Filip, and Miroslav Morcinek. 2016. "Waste and pollution in the ancient Roman Empire." *Journal of Landscape Ecology* 9 (3):33–49.

9. Wegmann, Edward. 1896. *The Water-supply of the City of New York. 1658–1895*, J. Wiley & Sons.

10. Scheader, Edward C. 1991. "The New York City water supply: Past, present and future." *Civil Engineering Practice* 6 (2):7–20.

11. Soll, David. 2013. *Empire of Water: An Environmental and Political History of the New York City Water Supply*, Cornell University Press.

12. Scheader, Edward C. 1991. "The New York City water supply: Past, present and future." *Civil Engineering Practice* 6 (2):7–20.

13. Gates, Alexander. 2023. *Polluted Earth: The Science of the Earth's Environment*, John Wiley & Sons.

14. Ceberio, Robert, and Ron Kase. 2015. *New Jersey Meadowlands: A History*, Arcadia Publishing.

15. Marshall, Stephen. 2004. "The Meadowlands before the commission: Three centuries of human use and alteration of the Newark and Hackensack Meadows." *Urban Habitats* 2 (1):4–27.

16. Ceberio, Robert, and Ron Kase. 2015. *New Jersey Meadowlands: A history*, Arcadia Publishing.

17. Gates, Alexander. 2023. *Polluted Earth: The Science of the Earth's Environment*, John Wiley & Sons.

18. Ceberio, Robert, and Ron Kase. 2015. *New Jersey Meadowlands: A History*, Arcadia Publishing.

19. Program, EPA Superfund Redevelopment. "Superfund Sites in Reuse in New Jersey." EPA, accessed November 4. https://www.epa.gov/superfund-redevelopment/superfund-sites-reuse-new-jersey.

20. Ceberio, Robert, and Ron Kase. 2015. *New Jersey Meadowlands: A History*, Arcadia Publishing.

21. Gates, Alexander E., and Robert P. Blauvelt. 2011. *Encyclopedia of Pollution*, Facts on File.

22. EPA. "Cuyahoga River AOC." EPA, last modified January 31, 2023, accessed November 4. https://www.epa.gov/great-lakes-aocs/cuyahoga-river-aoc.

23. Boissoneault, Lorraine. 2019. "The Cuyahoga River caught fire at least a dozen times, but no one cared until 1969." *Smithsonian Magazine*. Accessed November 4, 2023.

24. Gates, Alexander E., and Robert P. Blauvelt. 2011. *Encyclopedia of Pollution*, Facts on File.

25. Boissoneault, Lorraine. 2019. "The Cuyahoga River caught fire at least a dozen times, but no one cared until 1969." *Smithsonian Magazine*. Accessed November 4, 2023.

26. Blakemore, Erin. 2019. The shocking river fire that fueled the creation of the EPA. Accessed November 4, 2023. https://www.history.com/news/epa-earth-day-cleveland-cuyahoga-river-fire-clean-water-act.

27. Hines, N. William. 2013. "History of the 1972 Clean Water Act: The story behind how the 1972 act became the capstone on a decade of extraordinary

environmental reform." *George Washington Journal of Energy & Environmental Law* 4:80.

28. Gates, Alexander E., and Robert P. Blauvelt. 2011. *Encyclopedia of Pollution*, Facts on File.

29. Hines, N. William. 2013. "History of the 1972 Clean Water Act: The story behind how the 1972 act became the capstone on a decade of extraordinary environmental reform." *George Washington Journal of Energy & Environmental Law* 4:80.

30. Gates, Alexander. 2023. *Polluted Earth: The Science of the Earth's Environment*, John Wiley & Sons.

31. Area, USGS Water Resources Mission. 2019. Water quality in the nation's streams and rivers—Current conditions and long-term trends. Accessed November 4, 2023.

32. Leahy, P. Patrick, Joseph S. Rosenshein, and Debra S. Knopman. 1990. *Implementation Plan for the National Water-quality Assessment Program*. Vol. 90: Department of the Interior, U.S. Geological Survey.

33. Area, USGS Water Resources Mission. 2019. Water quality in the nation's streams and rivers—Current conditions and long-term trends. Accessed November 4, 2023.

34. Stets, Edward G., Lori A. Sprague, Gretchen P. Oelsner, Hank M. Johnson, Jennifer C. Murphy, Karen Ryberg, Aldo V. Vecchia, Robert E. Zuellig, James A. Falcone, and Melissa L. Riskin. 2020. "Landscape drivers of dynamic change in water quality of US rivers." *Environmental Science & Technology* 54 (7):4336–4343.

35. EPA. "National Rivers and Streams Assessment." EPA, last modified March 9, 2023, accessed November 4, 2023. https://www.epa.gov/national-aquatic-resource-surveys/nrsa.

36. Gates, Alexander. 2023. *Polluted Earth: The Science of the Earth's Environment*, John Wiley & Sons.

37. EPA. "Cuyahoga River AOC." EPA, last modified January 31, 2023, accessed November 4. https://www.epa.gov/great-lakes-aocs/cuyahoga-river-aoc.

38. Turner, R Eugene. 2021. "Declining bacteria, lead, and sulphate, and rising pH and oxygen in the lower Mississippi River." *Ambio* 50 (9):1731–1738.

39. Gates, Alexander E., and Robert P. Blauvelt. 2011. *Encyclopedia of Pollution*, Facts on File.

40. Stets, Edward G., Lori A. Sprague, Gretchen P. Oelsner, Hank M. Johnson, Jennifer C. Murphy, Karen Ryberg, Aldo V. Vecchia, Robert E. Zuellig, James A. Falcone, and Melissa L. Riskin. 2020. "Landscape drivers of dynamic change in water quality of US rivers." *Environmental Science & Technology* 54 (7):4336–4343.

41. Rakhmat, Dikanaya Tarahita and Muhammad Zulfikar. 2018. "Indonesia's Citarum: The world's most polluted river." *Asian Beat*. Accessed November 4, 2023.

42. Kerstens, Sjoerd. 2013. Downstream impacts of water pollution in the upper Citarum River, West Java, Indonesia: Economic assessment of interventions to improve water quality. The World Bank.

43. Rakhmat, Dikanaya Tarahita and Muhammad Zulfikar. 2018. "Indonesia's Citarum: The world's most polluted river." *Asian Beat*. Accessed November 4, 2023.

44. Shara, S., S. S. Moersidik, and T. E. B. Soesilo. 2021. "Potential health risks of heavy metals pollution in the downstream of Citarum River." IOP Conference Series: Earth and Environmental Science.

45. Sholeh, Muhammad, Pranoto Pranoto, Sri Budiastuti, and Sutarno Sutarno. 2018. "Analysis of Citarum River pollution indicator using chemical, physical, and bacteriological methods." AIP Conference Proceedings.

46. Shara, S., S. S. Moersidik, and T. E. B. Soesilo. 2021. "Potential health risks of heavy metals pollution in the downstream of Citarum River." IOP Conference Series: Earth and Environmental Science.

47. Kerstens, Sjoerd. 2013. Downstream impacts of water pollution in the upper Citarum River, West Java, Indonesia: Economic assessment of interventions to improve water quality. The World Bank.

48. Bukit, Nana Terangna. 1995. "Water quality conservation for the Citarum River in West Java." *Water Science and Technology* 31 (9):1–10.

49. Djuwita, Mitta Ratna, Djoko M. Hartono, Setyo S. Mursidik, and Tri Edhi Budhi Soesilo. 2021. "Pollution load allocation on water pollution control in the Citarum River." *Journal of Engineering & Technological Sciences* 53 (1).

50. Angelakis, Andreas N., Andrea G. Capodaglio, Cees W. Passchier, Mohammad Valipour, Jens Krasilnikoff, Vasileios A. Tzanakakis, Gül Sürmelihindi, Alper Baba, Rohitashw Kumar, and Benoît Haut. 2023. "Sustainability of water, sanitation, and hygiene: From prehistoric times to the present times and the future." *Water* 15 (8):1614.

51. Juuti, Petri S., Tapio Katko, and Heikkis Vuorinen. 2007. *Environmental History of Water*, IWA Publishing.

CHAPTER 8

1. Nelson, Gaylord. 1980. "Earth day'70: what it meant." *EPA Journal* 6 (4):6–38.

2. EPA. "Summary of the Comprehensive Environmental Response, Compensation, and Liability Act (Superfund)." EPA, Last Modified September 6, 2023, accessed November 5. https://www.epa.gov/laws-regulations/summary-comprehensive -environmental-response-compensation-and-liability-act.

3. Health, NYS Department of. 1978. Love Canal—Public Health Time Bomb. edited by Health. Albany, NY: NYS Department of Health.

4. Library, University of Buffalo. Love Canal: Timeline and Photos. Buffalo, NY.

5. Health, NYS Department of. 1978. Love Canal—Public Health Time Bomb. edited by Health. Albany, NY: NYS Department of Health.

6. Health, NYS Department of. 1978. Love Canal—Public Health Time Bomb. edited by Health. Albany, NY: NYS Department of Health.

7. EPA. "Superfund Site: LOVE CANAL, NIAGARA FALLS, NY Cleanup Activities." EPA, accessed November 5. https://cumulis.epa.gov/supercpad/SiteProfiles/index.cfm?fuseaction=second.cleanup&id=0201290.

8. Gates, Alexander E., and Robert P. Blauvelt. 2011. *Encyclopedia of Pollution*, Facts on File.

9. Beck, Eckardt C. 1979. "The Love Canal tragedy." *EPA Journal*, January 1979.

10. Health, NYS Department of. 1978. Love Canal—Public Health Time Bomb. edited by Health. Albany, NY: NYS Department of Health.

11. Beck, Eckardt C. 1979. "The Love Canal tragedy." *EPA Journal*, January 1979.

12. Library, University of Buffalo. Love Canal: Timeline and photos. Buffalo, NY.

13. Health, NYS Department of. 1978. Love Canal—Public Health Time Bomb. edited by Health. Albany, NY: NYS Department of Health.

14. Beck, Eckardt C. 1979. "The Love Canal tragedy." *EPA Journal*, January 1979.

15. Health, NYS Department of. 1978. Love Canal—Public health time bomb. edited by Health. Albany, NY: NYS Department of Health.

16. Beck, Eckardt C. 1979. "The Love Canal tragedy." *EPA Journal*, January 1979.

17. EPA. "Superfund Site: LOVE CANAL, NIAGARA FALLS, NY Cleanup Activities." EPA, accessed November 5. https://cumulis.epa.gov/supercpad/SiteProfiles/index.cfm?fuseaction=second.cleanup&id=0201290.

18. Gates, Alexander E., and Robert P. Blauvelt. 2011. *Encyclopedia of Pollution*, Facts on File.

19. EPA. "Superfund Site: LOVE CANAL, NIAGARA FALLS, NY Cleanup Activities." EPA, accessed November 5. https://cumulis.epa.gov/supercpad/SiteProfiles/index.cfm?fuseaction=second.cleanup&id=0201290.

20. Gates, Alexander E., and Robert P. Blauvelt. 2011. *Encyclopedia of Pollution*, Facts on File.

21. DeGroot, Rick. Fire in Chemical Waste Storage Facility Injures Dozens of Firefighters—Chemical Control Corporation Fire—Elizabeth, New Jersey—April 21, 1980. USFRA.org.

22. EPA. "Superfund Site: LOVE CANAL, NIAGARA FALLS, NY Cleanup Activities." EPA, accessed November 5. https://cumulis.epa.gov/supercpad/SiteProfiles/index.cfm?fuseaction=second.cleanup&id=0201290.

23. DeGroot, Rick. Fire in Chemical Waste Storage Facility Injures Dozens of Firefighters—Chemical Control Corporation Fire—Elizabeth, New Jersey - April 21, 1980. USFRA.org.

24. Nikfar, S., and N. Rahmani. 2014. Valley of the Drums. In *Encyclopedia of Toxicology* edited by Philip Wexler. New York, Academic Press.

25. EPA. "Superfund Site: A.L. TAYLOR (VALLEY OF DRUMS) BROOKS, KY." EPA, accessed November 5. https://cumulis.epa.gov/supercpad/cursites/csitinfo.cfm?id=0402072.

26. Nikfar, S., and N. Rahmani. 2014. Valley of the Drums. In *Encyclopedia of Toxicology* edited by Philip Wexler. New York: Academic Press.

27. EPA. "Superfund Site: A.L. TAYLOR (VALLEY OF DRUMS) BROOKS, KY." EPA, accessed November 5. https://cumulis.epa.gov/supercpad/cursites/csitinfo.cfm?id=0402072.

28. EPA. "Superfund Site: LOVE CANAL, NIAGARA FALLS, NY Cleanup Activities." EPA, accessed November 5. https://cumulis.epa.gov/supercpad/SiteProfiles/index.cfm?fuseaction=second.cleanup&id=0201290.

29. EPA. "Superfund Site: LOVE CANAL, NIAGARA FALLS, NY Cleanup Activities." EPA, accessed November 5. https://cumulis.epa.gov/supercpad/SiteProfiles/index.cfm?fuseaction=second.cleanup&id=0201290.

30. EPA. "Superfund Site: A.L. TAYLOR (VALLEY OF DRUMS) BROOKS, KY." EPA, accessed November 5. https://cumulis.epa.gov/supercpad/cursites/csitinfo.cfm?id=0402072.

31. EPA. "Summary of the Comprehensive Environmental Response, Compensation, and Liability Act (Superfund)." EPA, Last Modified September 6, 2023, accessed November 5. https://www.epa.gov/laws-regulations/summary-comprehensive-environmental-response-compensation-and-liability-act.

32. EPA. "Summary of the Comprehensive Environmental Response, Compensation, and Liability Act (Superfund)." EPA, Last Modified September 6, 2023, accessed November 5. https://www.epa.gov/laws-regulations/summary-comprehensive-environmental-response-compensation-and-liability-act.

33. Mintz, Joel A. 2011. "EPA enforcement of CERCLA: Historical overview and recent trends." *Southwestern University Law Review*, 41:645.

CHAPTER 9

1. U.S. Energy Information Administration. 2023. "U.S. energy facts explained." U.S. Energy Information Administration, last modified August 16, 2023, accessed November 6. https://www.eia.gov/energyexplained/us-energy-facts/.

2. Hill, Jason. 2022. "The sobering truth about corn ethanol." *Proceedings of the National Academy of Sciences* 119 (11):e2200997119.

3. Luo, Taotao. 2016. USDA: Energy efficiency of corn-ethanol production has improved significantly. Accessed November 6, 2023.

4. Diversity, Center for Biological. 2021. Eastern monarch butterfly population falls again. Center for Biological Diversity.

5. Patzek, Tad W., S.-M. Anti, R. Campos, K. W. Ha, J. Lee, B. Li, J. Padnick, and S.-A. Yee. 2005. "Ethanol from corn: clean renewable fuel for the future, or drain on our resources and pockets?" *Environment, Development and Sustainability* 7:319–336.

6. Scully, M. J., Norris, G. A., Alarcon Falconi, T. M. and MacIntosh, D. L. 2021. "Carbon intensity of corn ethanol in the United States: state of the science." *Environmental Research Letters* 16. doi: 10.1088/1748-9326/abde08.

7. Cheroennet, Nichanan, and Unchalee Suwanmanee. 2017. "Net energy gain and water footprint of corn ethanol production in Thailand." *Energy Procedia* 118:15–20.

8. Energy.gov. 2023. Wind market reports: 2023 Edition. Office of Energy Efficiency & Renewable Energy.

9. U.S. Energy Information Administration. 2023. "U.S. energy facts explained." U.S. Energy Information Administration, last modified August 16, 2023, accessed November 6. https://www.eia.gov/energyexplained/us-energy-facts/.

10. Schippers, Peter, Ralph Buij, Alex Schotman, Jana Verboom, Henk van der Jeugd, and Eelke Jongejans. 2020. "Mortality limits used in wind energy impact

assessment underestimate impacts of wind farms on bird populations." *Ecology and Evolution* 10 (13):6274–6287.

11. International, Bat Conservation. 2023. "State of the bats, North America 2023." Bat Conservation International, accessed November 6. https://digital.batcon.org/state-of-the-bats-report/2023-report/.

12. Rodhouse, Thomas J., Rogelio M. Rodriguez, Katharine M. Banner, Patricia C. Ormsbee, Jenny Barnett, and Kathryn M. Irvine. 2019. "Evidence of region-wide bat population decline from long-term monitoring and Bayesian occupancy models with empirically informed priors." *Ecology and Evolution* 9 (19):11078–11088.

13. Energy.gov. 2017. "How Do Wind Turbines Survive Severe Storms?" U.S. Department of Energy, last modified June 20, 2017, accessed November 5. https://www.energy.gov/eere/articles/how-do-wind-turbines-survive-severe-storms.

14. Szilvia Doczi Tormey and Jennifer Chen. 2019. U.S. regulatory innovation to boost power system flexibility and prepare for ramp up of wind and solar. Accessed November 7, 2023. https://www.iea.org/commentaries/us-regulatory-innovation-to-boost-power-system-flexibility-and-prepare-for-ramp-up-of-wind-and-solar.

15. Energy.gov. 2017. "How do wind turbines survive severe storms?" U.S. Department of Energy, last modified June 20, 2017, accessed November 5. https://www.energy.gov/eere/articles/how-do-wind-turbines-survive-severe-storms.

16. Musial, Walter, Paul Spitsen, Philipp Beiter, Patrick Duffy, Melinda Marquis, Rob Hammond, and Matt Shields. 2023. Offshore wind market report: 2022 Edition. Washington, DC.

17. Hartman, L. 2018. Wind turbines in extreme weather: Solutions for hurricane resiliency. Accessed November 5, 2023.

18. Energy.gov. 2023. Wind market reports: 2023 edition. Office of Energy Efficiency & Renewable Energy.

19. Gates, Alexander. 2023. *Polluted Earth: The Science of the Earth's Environment*, John Wiley & Sons.

20. Energy.gov. 2023. Wind market reports: 2023 edition. Office of Energy Efficiency & Renewable Energy.

21. Energy.gov. 2023. Wind market reports: 2023 edition. Office of Energy Efficiency & Renewable Energy.

22. European Photovoltaic Industry Association. 2011. Solar voltaics: Competing in the energy sector.

23. U.S. Energy Information Administration. 2023. "U.S. energy facts explained." U.S. Energy Information Administration, last modified August 16, 2023, accessed November 6. https://www.eia.gov/energyexplained/us-energy-facts/.

24. Sharma, R. 2019. "Effect of obliquity of incident light on the performance of silicon solar cells." *Heliyon* 5 (7).

25. Sharma, R. 2019. "Effect of obliquity of incident light on the performance of silicon solar cells." *Heliyon* 5 (7).

26. Energy.gov. "How Does Solar Work?" U.S. Department of Energy, accessed November 5. https://www.energy.gov/eere/solar/how-does-solar-work.

27. Masmitjà, Gerard, Eloi Ros, Rosa Almache-Hernández, Benjamín Pusay, Isidro Martín, Cristóbal Voz, Edgardo Saucedo, Joaquim Puigdollers, and Pablo Ortega.

2022. "Interdigitated back-contacted crystalline silicon solar cells fully manufactured with atomic layer deposited selective contacts." *Solar Energy Materials and Solar Cells* 240:111731.

28. U.S. Energy Information Administration. 2021. Solar generation was 3% of U.S. electricity in 2020, but we project it will be 20% by 2050. *Today in Energy* (November 2021). Accessed November 5, 2023.

29. U.S. Energy Information Administration. 2023. "U.S. energy facts explained." U.S. Energy Information Administration, last modified August 16, 2023, accessed November 6. https://www.eia.gov/energyexplained/us-energy-facts/.

30. U.S. Energy Information Administration. 2023. "U.S. energy facts explained." U.S. Energy Information Administration, last modified August 16, 2023, accessed November 6. https://www.eia.gov/energyexplained/us-energy-facts/.

31. Energy.gov. "Hydropower basics." U.S. Department of Energy, accessed November 5. https://www.energy.gov/eere/water/hydropower-basics.

32. USGS. 2018. "Hydroelectric power: How it works." USGS, accessed November 5. https://www.usgs.gov/special-topics/water-science-school/science/hydroelectric -power-how-it-works#science.

33. USDOE, Alternative Fuels Data Center. 2023. Electric Vehicle Benefits and Considerations. Accessed November 6, 2023.

34. Board, California Air Resources. 2022. California moves to accelerate to 100% new zero-emission vehicle sales by 2035. 2023 (November 6).

35. USDOE, Alternative Fuels Data Center. 2023. Electric Vehicle Benefits and Considerations. Accessed November 6, 2023.

36. Shine, I. 2022. The world needs 2 billion electric vehicles to get to net zero. But is there enough lithium to make all the batteries? Accessed November 5, 2023.

37. IEA. 2023. Global EV Outlook 2023.

38. Shine, I. 2022. The world needs 2 billion electric vehicles to get to net zero. But is there enough lithium to make all the batteries? Accessed November 5, 2023.

39. Russo, Danny. 2023. EV battery replacement cost. *Consumer Affairs*. Accessed November 6, 2023.

40. KBB, Kelley Blue Book. 2022. New-Vehicle Prices Set a Record in June, According to Kelley Blue Book, as Luxury Share Hits New High.

41. Bureau, U.S. Census. 2022. Nation's urban and rural populations shift following 2020 census. Washington, DC: U.S. Census Bureau.

42. UN, United Nations. 2018. 68% of the world population projected to live in urban areas by 2050, says UN. Accessed November 6, 2023.

43. MacDonald, A. E., Clack, C. T., Alexander, A., Dunbar, A., Wilczak, J. and Xie, Y. 2016. "Future cost-competitive electricity systems and their impact on US CO2 emissions." *Nature Climate Change* 6 (5):526–31. doi: 10.1038/nclimate2921.

44. Department, Statistica Research. 2023. "Number of power outages in the United States between 2008 and 2017, by select state." Statistica, last modified January 25, 2023, accessed November 5. https://www.statista.com/statistics/1078354 /electricity-blackouts-by-state/.

45. Board, California Air Resources. 2022. California moves to accelerate to 100% new zero-emission vehicle sales by 2035. 2023 (November 6).

46. Lark, Tyler J., Nathan P. Hendricks, Aaron Smith, Nicholas Pates, Seth A. Spawn-Lee, Matthew Bougie, Eric G. Booth, Christopher J. Kucharik, and Holly K. Gibbs. 2022. "Environmental outcomes of the US renewable fuel standard." *Proceedings of the National Academy of Sciences* 119 (9):e2101084119.

47. Szilvia Doczi Tormey and Jennifer Chen. 2019. U.S. regulatory innovation to boost power system flexibility and prepare for ramp up of wind and solar. Accessed November 7, 2023.

CHAPTER 10

1. U.S. Energy Information Administration. "U.S. energy facts explained." U.S. Energy Information Administration, accessed November 11. https://www.eia.gov/energyexplained/us-energy-facts/.

2. Morse, Andrew Turgeon and Elizabeth. Geothermal energy. Accessed November 11, 2023.[AU: Is there a link missing here? What publication is this source from?] https://education.nationalgeographic.org/resource/geothermal-energy/

3. Dickson, Mary H., and Mario Fanelli. 2013. "Geothermal energy: Utilization and technology," 205, p. 2003, UNESCO/Routledge.

4. Mims, Christopher. 2008. "One hot island: Iceland's renewable geothermal power." *Scientific American*, October 20, 2008.

5. Fernandez, Lucia. 2023. Geothermal energy generation worldwide in 2021, by country. Accessed November 11, 2023. https://www.statista.com/statistics/514488/geothermal-generation-worldwide-by-key-country/

6. Bank, The World. 2017. Geothermal. 11. Accessed November 11, 2023. https://www.worldbank.org/en/results/2017/12/01/geothermal

7. Office, Hawaii State Energy. "Renewable energy resources." State of Hawaii, accessed November 11. https://energy.hawaii.gov/what-we-do/energy-landscape/renewable-energy-resources/.

8. Dickson, Mary H., and Mario Fanelli. 2013. "Geothermal energy: Utilization and technology."

9. Narsilio, Guillermo Andres, and Lu Aye. 2018. "Shallow geothermal energy: an emerging technology." *Low Carbon Energy Supply: Trends, Technology, Management*:387–411.

10. Narsilio, Guillermo Andres, and Lu Aye. 2018. "Shallow geothermal energy: an emerging technology." *Low Carbon Energy Supply: Trends, Technology, Management*:387–411.

11. Bioenergy, IEA. 2021. Implementation of bioenergy in Brazil – 2021 update. *Country Reports*. Accessed November 11, 2023.

12. Dahiya, Anju. 2014. *Bioenergy: Biomass to Biofuels*, Academic Press.

13. John Sheehan, Vince Camobreco, James Duffield, Michael Graboski, Housein Shapouri. 1998. An overview of biodiesel and petroleum diesel life cycles. U.S. Department of Agriculture and U.S. Department of Energy.

14. Energy, Farm. 2019. "Soybeans for biodiesel production." Farm Energy, last modified April 3, 2019, accessed November 11. https://farm-energy.extension.org/soybeans-for-biodiesel-production/.

15. Charles, Dan. 2018. Turning soybeans into diesel fuel is costing us billions. https://www.npr.org/sections/thesalt/2018/01/16/577649838/turning-soybeans-into-diesel-fuel-is-costing-us-billions

16. Statistica. 2023. "Biodiesel consumption in the U.S. 2007–2022." Statistica Research Department, accessed November 11. https://www.statista.com/statistics/509894/consumption-volume-of-biodiesel-in-the-us/.

17. Waseem, Hafiza Hafza, Asma El Zerey-Belaskri, Farwa Nadeem, and Iqra Yaqoob. 2016. "The downside of biodiesel fuel—A review." *International Journal of Chemical and Biochemical Sciences* 9:97–106.

18. John Sheehan, Vince Camobreco, James Duffield, Michael Graboski, Housein Shapouri. 1998. An overview of biodiesel and petroleum diesel life cycles. U.S. Department of Agriculture and U.S. Department of Energy.

19. El-Sheekh, Mostafa, and Abd El-Fatah Abomohra. 2021. *Handbook of Algal Biofuels: Aspects of Cultivation, Conversion, and Biorefinery*, Elsevier.

20. Körner, Sabine, Jan E. Vermaat, and Siemen Veenstra. 2003. "The capacity of duckweed to treat wastewater: Ecological considerations for a sound design." *Journal of Environmental Quality* 32 (5):1583–1590.

21. Ananyev, Gennady, Colin Gates, Aaron Kaplan, and G. Charles Dismukes. 2017. "Photosystem II-cyclic electron flow powers exceptional photoprotection and record growth in the microalga Chlorella ohadii." *Biochimica et Biophysica Acta (BBA)—Bioenergetics* 1858 (11):873–83.

22. Rodionova, Margarita V., Roshan Sharma Poudyal, Indira Tiwari, Roman A. Voloshin, Sergei K. Zharmukhamedov, Hong Gil Nam, Bolatkhan K Zayadan, Barry D. Bruce, Harvey J. M. Hou, and Suleyman I. Allakhverdiev. 2017. "Bio-fuel production: Challenges and opportunities." *International Journal of Hydrogen Energy* 42 (12):8450–8461.

23. Rodionova, Margarita V., Roshan Sharma Poudyal, Indira Tiwari, Roman A. Voloshin, Sergei K. Zharmukhamedov, Hong Gil Nam, Bolatkhan K. Zayadan, Barry D. Bruce, Harvey J. M. Hou, and Suleyman I. Allakhverdiev. 2017. "Bio-fuel production: Challenges and opportunities." *International Journal of Hydrogen Energy* 42 (12):8450–8461.

24. El-Sheekh, Mostafa, and Abd El-Fatah Abomohra. 2021. *Handbook of Algal Biofuels: Aspects of Cultivation, Conversion, and Biorefinery*, Elsevier.

25. Dahiya, Anju. 2014. *Bioenergy: Biomass to Biofuels*, Academic Press.

26. John Sheehan, Vince Camobreco, James Duffield, Michael Graboski, Housein Shapouri. 1998. An overview of biodiesel and petroleum diesel life cycles. U.S. Department of Agriculture and U.S. Department of Energy.

27. Dahiya, Anju. 2014. *Bioenergy: Biomass to Biofuels*, Academic Press.

28. U.S. Energy Information Administration. 2023. "Nuclear power plants generated 68% of France's electricity in 2021." EIA, last modified January 23, 2023, accessed November 11. https://www.eia.gov/todayinenergy/detail.php?id=55259.

29. Galindo, Andrea. 2022. What is nuclear energy? The science of nuclear power. Accessed November 11, 2023. https://www.iaea.org/newscenter/news/what-is-nuclear-energy-the-science-of-nuclear-power

30. Breeze, Paul. 2016. *Nuclear Power*, Academic Press.

31. Breeze, Paul. 2016. *Nuclear Power*, Academic Press.

32. Galindo, Andrea. 2022. What is nuclear energy? The science of nuclear power.

33. Breeze, Paul. 2016. *Nuclear Power*, Academic Press.

34. Association, World Nuclear. "Supply of uranium." World Nuclear Association, last modified August 2023, accessed November 11. https://world-nuclear.org/information-library/nuclear-fuel-cycle/uranium-resources/supply-of-uranium.aspx.

35. Association, World Nuclear. "Supply of uranium." World Nuclear Association, last modified August 2023, accessed November 11. https://world-nuclear.org/information-library/nuclear-fuel-cycle/uranium-resources/supply-of-uranium.aspx.

36. Gates, Alexander. 2023. *Polluted Earth: The Science of the Earth's Environment*, John Wiley & Sons.

37. Association, World Nuclear. "Supply of uranium." World Nuclear Association, last modified August 2023, accessed November 11. https://world-nuclear.org/information-library/nuclear-fuel-cycle/uranium-resources/supply-of-uranium.aspx.

38. Breeze, Paul. 2016. *Nuclear Power*, Academic Press.

39. EPA. 2017. "Carbon capture and sequestration overview." EPA, last modified January 19, 2017, accessed November 11. https://19january2017snapshot.epa.gov/climatechange/carbon-dioxide-capture-and-sequestration-overview_.html.

40. Congressional Research Service, Report 2022. Carbon capture and sequestration (CCS) in the United States.

41. Energy.gov. 2017. Petra Nova—W.A. Parish Project.

42. Energy, U.S. Dept of. "Carbon storage FAQs, National Energy Technology Laboratory." National Energy Technology Laboratory, accessed November 11. https://www.netl.doe.gov/carbon-management/carbon-storage/faqs/carbon-storage-faqs.

43. Energy, U.S. Dept of. "Ethanol laws and incentives in federal." Alternative Energy Data Center, accessed November 11. https://afdc.energy.gov/fuels/laws/ETH?state=US.

44. Casey-Lefkowitz, Susan and Sujatha Bergen. 2023. Congress should follow Biden's lead on fossil fuel subsidies. Accessed November 11, 2023. https://www.nrdc.org/bio/sujatha-bergen/congress-should-follow-bidens-lead-fossil-fuel-subsidies.

45. Skene, Jennifer. 2019. The issue with tissue: How Americans are flushing forests down the toilet. https://www.nrdc.org/sites/default/files/issue-tissue-how-americans-are-flushing-forests-down-toilet-report.pdf.

CHAPTER 11

1. Liang, Youye. 2010. "A long lasting and extensive drought event over China in 1876–1878." *Advances in Climate Change Research* 1 (2):91–99.

2. WWAP (UNESCO World Water Assessment Programme). https://www.unesdoc .unesco.org/ark:/48223/pf0000367306. 2019. *The United Nations World Water Development Report 2019: Leaving No One Behind.* Paris: UNESCO.

3. Roser, Hannah Ritchie and Max. 2018. Water use and stress. Accessed November 11, 2023. https://ourworldindata.org/water-use-stress.

4. Micklin, Philip. 2007. "The Aral Sea disaster." *Annual Review of Earth and Planetary Sciences* 35:47–72.

5. Gates, Alexander. 2023. *Polluted Earth: The Science of the Earth's Environment,* John Wiley & Sons.

6. Gates, Alexander. 2023. *Polluted Earth: The Science of the Earth's Environment,* John Wiley & Sons.

7. Gates, Alexander. 2023. *Polluted Earth: The Science of the Earth's Environment,* John Wiley & Sons.

8. Little, Jane Braxton. 2009. "The Ogallala Aquifer: Saving a vital US water source." *Scientific American* 1.

9. Gates, Alexander. 2023. *Polluted Earth: The Science of the Earth's Environment,* John Wiley & Sons.

10. Little, Jane Braxton. 2009. "The Ogallala Aquifer: Saving a vital US water source." *Scientific American* 1.

11. Gates, Alexander. 2023. *Polluted Earth: The Science of the Earth's Environment,* John Wiley & Sons.

12. Ramaswamy, Sonny. 2016. "Reversing pollinator decline is key to feeding the future." U.S. Department of Agriculture. Retrieved from https://www.usda.gov/ media/blog/2016/06/24/reversing-pollinator-decline-key-feedingfuture.

13. Burd, Kelsey Kopec and Lori Ann. 2017. Pollinators in peril. https://www .biologicaldiversity.org/campaigns/native_pollinators/pdfs/Pollinators_in_Peril.pdf.

14. Smith, Kristine M., Elizabeth H. Loh, Melinda K. Rostal, Carlos M. Zambrana-Torrelio, Luciana Mendiola, and Peter Daszak. 2013. "Pathogens, pests, and economics: Drivers of honey bee colony declines and losses." *EcoHealth* 10:434–45.

15. Pettis, Jeffery S., and Keith S. Delaplane. 2010. "Coordinated responses to honey bee decline in the USA." *Apidologie* 41 (3):256–63.

16. Janousek, William M., Margaret R. Douglas, Syd Cannings, Marion A. Clément, Casey M. Delphia, Jeffrey G. Everett, Richard G. Hatfield, Douglas A. Keinath, Jonathan B. Uhuad Koch, and Lindsie M. McCabe. 2023. "Recent and future declines of a historically widespread pollinator linked to climate, land cover, and pesticides." *Proceedings of the National Academy of Sciences* 120 (5):e2211223120.

17. Sánchez-Bayo, Francisco, and Kris A. G. Wyckhuys. 2019. "Worldwide decline of the entomofauna: A review of its drivers." *Biological Conservation* 232:8–27.

18. Gates, Alexander E., and Robert P. Blauvelt. 2011. *Encyclopedia of Pollution,* Facts on File.

19. Gates, Alexander E., and Robert P. Blauvelt. 2011. *Encyclopedia of Pollution,* Facts on File.

20. Smith, Kristine M., Elizabeth H. Loh, Melinda K. Rostal, Carlos M. Zambrana-Torrelio, Luciana Mendiola, and Peter Daszak. 2013. "Pathogens,

pests, and economics: Drivers of honey bee colony declines and losses." *EcoHealth* 10:434–45.

21. Smith, Kristine M., Elizabeth H. Loh, Melinda K. Rostal, Carlos M. Zambrana-Torrelio, Luciana Mendiola, and Peter Daszak. 2013. "Pathogens, pests, and economics: Drivers of honey bee colony declines and losses." *EcoHealth* 10:434–45.

22. Wagner, David L., Eliza M. Grames, Matthew L. Forister, May R. Berenbaum, and David Stopak. 2021. "Insect decline in the Anthropocene: Death by a thousand cuts." *Proceedings of the National Academy of Sciences* 118 (2):e2023989118.

23. Wagner, David L., Eliza M. Grames, Matthew L. Forister, May R. Berenbaum, and David Stopak. 2021. "Insect decline in the Anthropocene: Death by a thousand cuts." *Proceedings of the National Academy of Sciences* 118 (2):e2023989118.

24. Zattara, Eduardo E., and Marcelo A. Aizen. 2021. "Worldwide occurrence records suggest a global decline in bee species richness." *One Earth* 4 (1):114–23.

25. Rodhouse, Thomas J., Rogelio M. Rodriguez, Katharine M. Banner, Patricia C. Ormsbee, Jenny Barnett, and Kathryn M. Irvine. 2019. "Evidence of region-wide bat population decline from long-term monitoring and Bayesian occupancy models with empirically informed priors." *Ecology and Evolution* 9 (19):11078–11088.

26. Seibold, Sebastian, Martin M. Gossner, Nadja K. Simons, Nico Blüthgen, Jörg Müller, Didem Ambarlı, Christian Ammer, Jürgen Bauhus, Markus Fischer, and Jan C. Habel. 2019. "Arthropod decline in grasslands and forests is associated with landscape-level drivers." *Nature* 574 (7780):671–74.

27. Rosenberg, Kenneth V., Adriaan M. Dokter, Peter J. Blancher, John R. Sauer, Adam C. Smith, Paul A. Smith, Jessica C. Stanton, Arvind Panjabi, Laura Helft, and Michael Parr. 2019. "Decline of the North American avifauna." *Science* 366 (6461):120–24.

28. Geographic, National. Dead Zone. National Geographic. https://education.nationalgeographic.org/resource/dead-zone/.

29. Howard, Jenny. 2019. Dead zones, explained. *National Geographic*. Accessed November 11, 2023. https://www.nationalgeographic.com/environment/article/dead-zones?loggedin=true&rnd=1706792199321.

30. Diaz, Robert J., and Rutger Rosenberg. 2008. "Spreading dead zones and consequences for marine ecosystems." *Science* 321 (5891):926–29.

31. Costello, Christopher, Ling Cao, Stefan Gelcich, Miguel Á. Cisneros-Mata, Christopher M. Free, Halley E. Froehlich, Christopher D. Golden, Gakushi Ishimura, Jason Maier, and Ilan Macadam-Somer. 2020. "The future of food from the sea." *Nature* 588 (7836):95–100.

32. Gates, Alexander. 2023. *Polluted Earth: The Science of the Earth's Environment*, John Wiley & Sons.

33. Antweiler, Ronald C., Donald A. Goolsby, and Howard E. Taylor. 1996. "Nutrients in the Mississippi river." *U.S. Geological Survey Circular USGS Circular*:73–86.

34. EPA. 2011. "Nitrogen and phosphorus pollution in the Mississippi River basin: Findings of the wadeable streams assessment." EPA, last modified December 2011, accessed November 11. https://www.epa.gov/sites/default/files/2015-03/documents/epa-marb-fact-sheet-112911_508.pdf.

35. EPA. 2011. "Nitrogen and phosphorus pollution in the Mississippi River basin: Findings of the wadeable streams assessment." EPA, last modified December 2011, accessed November 11. https://www.epa.gov/sites/default/files/2015-03/documents/epa-marb-fact-sheet-112911_508.pdf.

36. Owen, James. 2012. "World's largest dead zone suffocating sea." *National Geographic News* 5.

37. Carstensen, Jacob, and Daniel J. Conley. 2019. "Baltic Sea hypoxia takes many shapes and sizes." *Limnology and Oceanography Bulletin* 28 (4):125–29.

CHAPTER 12

1. Ying Cui, Lee R. Kump. 2015. "Global warming and the end-Permian extinction event: Proxy and modeling perspectives." *Earth-Science Reviews* 149:5–22. doi: https://doi.org/10.1016/j.earscirev.2014.04.007.

2. Gates, Alexander. 2023. *Polluted Earth: The Science of the Earth's Environment*, John Wiley & Sons.

3. Wu, Yuyang, Daoliang Chu, Jinnan Tong, Haijun Song, Jacopo Dal Corso, Paul B. Wignall, Huyue Song, Yong Du, and Ying Cui. 2021. "Six-fold increase of atmospheric p CO_2 during the Permian–Triassic mass extinction." *Nature Communications* 12 (1):2137.

4. Benton, M. J. 2018. "Hyperthermal-driven mass extinctions: Killing models during the Permian–Triassic mass extinction." *Philosophical Transactions of the Royal Society A*. doi: 10.1098/rsta.2017.0076.

5. Gliwa, Jana, Michael Wiedenbeck, Martin Schobben, Clemenz V. Ullmann, Wolfgang Kiessling, Abbas Ghaderi, Ulrich Struck, and Dieter Korn. 2022. "Gradual warming prior to the end-Permian mass extinction." *Palaeontology* 65 (5):e12621.

6. Gates, Alexander. 2023. *Polluted Earth: The Science of the Earth's Environment*, John Wiley & Sons.

7. Wu, Yuyang, Daoliang Chu, Jinnan Tong, Haijun Song, Jacopo Dal Corso, Paul B. Wignall, Huyue Song, Yong Du, and Ying Cui. 2021. "Six-fold increase of atmospheric p CO_2 during the Permian–Triassic mass extinction." *Nature Communications* 12 (1):2137.

8. Ying Cui, Lee R. Kump. 2015. "Global warming and the end-Permian extinction event: Proxy and modeling perspectives." *Earth-Science Reviews* 149:5–22. doi: https://doi.org/10.1016/j.earscirev.2014.04.007.

9. Gates, Alexander. 2022. *Earth's Fury: The Science of Natural Disasters*, John Wiley & Sons.

10. Benton, M. J. 2018. "Hyperthermal-driven mass extinctions: Killing models during the Permian–Triassic mass extinction." *Philosophical Transactions of the Royal Society A*. doi: 10.1098/rsta.2017.0076.

11. Song, Haijun, David B. Kemp, Li Tian, Daoliang Chu, Huyue Song, and Xu Dai. 2021. "Thresholds of temperature change for mass extinctions." *Nature Communications* 12 (1):4694.

12. Lindsey, Rebecca and LuAnn Dahlman. 2023. Climate Change: Global Temperature. Accessed November 14, 2023. https://www.climate.gov/news-features/understanding-climate/climate-change-global-temperature.

13. U.S. Global Change Research Program. 2017. Climate science special Report: Fourth national climate assessment, edited by D. J. Wuebbles, D. W. Fahey, K. A. Hibbard, D. J. Dokken, B. C. Stewart, and T. K. Maycock Washington, DC: U.S. Global Change Research Program.

14. Herring, David. 2012. Climate Change: Global Temperature Projections. Accessed November 14, 2023. https://www.climate.gov/news-features/understanding-climate/climate-change-global-temperature-projections.

15. U.S. Global Change Research Program. 2017. Climate science special report: Fourth national climate assessment, edited by D. J. Wuebbles, D. W. Fahey, K. A. Hibbard, D. J. Dokken, B. C. Stewart, and T. K. Maycock Washington, DC: U.S. Global Change Research Program.

Bibliography

Agency for Toxic Substances and Disease Registry, ATSDR. 2020. "Toxicological Profile for Lead." Accessed October 26. https://www.atsdr.cdc.gov/ToxProfiles/tp13.pdf.

Agency for Toxic Substances and Disease Registry, ATSDR. 2022. "ATSDR's Substance Priority List." Accessed October 26. https://www.atsdr.cdc.gov/spl/index.html.

Agency for Toxic Substances and Disease Registry (ASTDR). 2022. "Toxicological Profile for DDT, DDE, and DDD." U.S. Centers for Disease Control, accessed October 26. https://www.atsdr.cdc.gov/toxprofiles/tp35.pdf.

American Chemical Society, ACS. 2017. "Chlorofluorocarbons and Ozone Depletion." Accessed October 28. http://www.acs.org/content/acs/en/education/whatischemistry/landmarks/cfcs-ozone.html.

Ananyev, Gennady, Colin Gates, Aaron Kaplan, and G. Charles Dismukes. 2017. "Photosystem II-cyclic electron flow powers exceptional photoprotection and record growth in the microalga Chlorella ohadii." *Biochimica et Biophysica Acta (BBA)—Bioenergetics* 1858 (11):873–83.

Angelakis, Andreas N., Andrea G. Capodaglio, Cees W. Passchier, Mohammad Valipour, Jens Krasilnikoff, Vasileios A. Tzanakakis, Gül Sürmelihindi, Alper Baba, Rohitashw Kumar, and Benoît Haut. 2023. "Sustainability of Water, Sanitation, and Hygiene: From Prehistoric Times to the Present Times and the Future." *Water* 15 (8):1614.

Antweiler, Ronald C., Donald A. Goolsby, and Howard E. Taylor. 1996. "Nutrients in the Mississippi river." *U.S. Geological Survey Circular USGS Circ*:73–86.

ATSDR. 1998. "Toxicological Profile for Sulfur Dioxide." U.S. Dept of Health and Human Services, accessed October 30. https://www.atsdr.cdc.gov/toxprofiles/tp116.pdf.

Bat Conservation International. 2023. "State of the Bats, North America 2023." Bat Conservation International, accessed November 6. https://digital.batcon.org/state-of-the-bats-report/2023-report/.

Berry-Caban, Cristobal S. 2011. "DDT and Silent Spring: Fifty years after." *Journal of Military and Veteran Health* 19:19–24.

Benton, Michael J., and Richard J. Twitchett. 2003. "How to kill (almost) all life: The end-Permian extinction event." *Trends in Ecology & Evolution* 18 (7):358–65.

Beck, Eckardt C. 1979. "The Love Canal Tragedy." *EPA Journal*, January 1979.

Benton, M. J. 2018. "Hyperthermal-driven mass extinctions: Killing models during the Permian–Triassic mass extinction." *Philosophical Transactions of the Royal Society A.* doi: 10.1098/rsta.2017.0076.

Blakemore, Erin. 2016. "The Ozone Hole Was Super Scary, So What Happened to It?" *Smithsonian Magazine*.

Blakemore, Erin. 2019. The Shocking River Fire That Fueled the Creation of the EPA. Accessed November 4, 2023. https://www.history.com/news/epa-earth-day -cleveland-cuyahoga-river-fire-clean-water-act.

Boissoneault, Lorraine. 2019. The Cuyahoga River Caught Fire at Least a Dozen Times, but No One Cared Until 1969. *Smithsonian Magazine*. Accessed November 4, 2023.

Breeze, Paul. 2016. *Nuclear Power*, Academic Press.

Brimblecombe, P. 2012. *The Big Smoke: A History of Air Pollution in London since Medieval Times*, Taylor & Francis.

Buchholtz, K. 2020. "2010s see Record Number of Storms in the Philippines." Statista.com, accessed October 24. https://www.statista.com/chart/23638/storms -storm-deaths-in-the-philippines/.

Bukit, Nana Terangna. 1995. "Water quality conservation for the Citarum River in West Java." *Water Science and Technology* 31 (9):1–10.

Burd, Kelsey Kopec and Lori Ann. 2017. Pollinators in Peril: A Systematic Status Review of North American and Hawaiian Native Bees. Center for Biological Diversity. https://www.biologicaldiversity.org/campaigns/native_pollinators/pdfs/ Pollinators_in_Peril.pdf.

Burger, Joanna. 1997. *Oil Spills*. New Brunswick, NJ: Rutgers University Press.

Cacciottolo, Bethan Bell and Mario. 2017. Torrey Canyon oil spill: The day the sea turned black. BBC, March 17. https://www.bbc.com/news/uk-england-39223308.

Calabrese, Edward J. 1982. "Human breast milk contamination in the United States and Canada by chlorinated hydrocarbon insecticides and industrial pollut- ants: Current status." *Journal of the American College of Toxicology* 1 (3):91–98.

California Air Resources Board. 2022. California moves to accelerate to 100% new zero-emission vehicle sales by 2035. 2023 (November 6).

Carson, Rachel. 2009. *Silent Spring*, first ed. 1962. Los Angeles: Getty Publications.

Carstensen, Jacob, and Daniel J. Conley. 2019. "Baltic Sea hypoxia takes many shapes and sizes." *Limnology and Oceanography Bulletin* 28 (4):125–129.

Casey-Lefkowitz, Susan and Sujatha Bergen. 2023. Congress Should Follow Biden's Lead on Fossil Fuel Subsidies. Accessed November 11, 2023. https://www.nrdc .org/bio/sujatha-bergen/congress-should-follow-bidens-lead-fossil-fuel-subsidies.

CDC, Centers for Disease Control and Prevention. 2016. "Flint Water Crisis." Last modified October 24, 2023. https://www.cdc.gov/nceh/casper/pdf-html/flint_water _crisis_pdf.html.

Ceberio, Robert, and Ron Kase. 2015. *New Jersey Meadowlands: A History*, Arcadia Publishing.

Cedre. "Atlantic Empress/Aegean Captain." Cedre, last modified August 2, 2007, accessed October 31. https://wwz.cedre.fr/en/Resources/Spills/Spills/Atlantic-Empress-Aegean-Captain.

Center for Biological Diversity. 2021. Eastern Monarch Butterfly Population Falls Again. Center for Biological Diversity.

Charles, Dan. 2018. Turning Soybeans into Diesel Fuel Is Costing Us Billions, NPR, January 16. https://www.npr.org/sections/thesalt/2018/01/16/577649838/turning-soybeans-into-diesel-fuel-is-costing-us-billions.

Cheroennet, Nichanan, and Unchalee Suwanmanee. 2017. "Net energy gain and water footprint of corn ethanol production in Thailand." *Energy Procedia* 118:15–20.

Commission, Alaska Oil Spill. 1990. Spill: The wreck of the Exxon Valdez, Final Report. Alaska: State of Alaska.

Copernicus. 2022. "River Discharge." European State of the Climate, accessed October 25. https://climate.copernicus.eu/esotc/2022/river-discharge.

Costello, Christopher, Ling Cao, Stefan Gelcich, Miguel Á. Cisneros-Mata, Christopher M. Free, Halley E. Froehlich, Christopher D. Golden, Gakushi Ishimura, Jason Maier, and Ilan Macadam-Somer. 2020. "The future of food from the sea." *Nature* 588 (7836):95–100.

Cui, Ying, and Lee R. Kump. 2015. "Global warming and the end-Permian extinction event: Proxy and modeling perspectives." *Earth-Science Reviews* 149:5–22. doi: https://doi.org/10.1016/j.earscirev.2014.04.007.

Dahiya, Anju. 2014. *Bioenergy: Biomass to biofuels*, Academic Press.

DeGroot, Rick. Fire in Chemical Waste Storage Facility Injures Dozens of Firefighters—Chemical Control Corporation Fire—Elizabeth, New Jersey, April 21, 1980. USFRA.org.

Delile, Hugo, Janne Blichert-Toft, Jean-Philippe Goiran, Simon Keay, and Francis Albarède. 2014. "Lead in ancient Rome's city waters." *Proceedings of the National Academy of Sciences* 111 (18):6594–99.

Deming, David. 2020. "The aqueducts and water supply of Ancient Rome." *Ground water* 58 (1):152.

Denchak, M. 2018. "Flint Water Crisis: Everything You Need to Know." Natural Resources Defense Council NDRC. https://www.nrdc.org/stories/flint-water-crisis-everything-you-need-know.

Diaz, Robert J., and Rutger Rosenberg. 2008. "Spreading dead zones and consequences for marine ecosystems." *Science* 321 (5891):926–29.

Dickson, Mary H., and Mario Fanelli. 2013. "Geothermal energy: utilization and technology."

Dillon, J. C., G. B. Martin, and H. T. O'Brien. 1981. "Pesticide residues in human milk." *Food and Cosmetics Toxicology* 19:437–42.

Djuwita, Mitta Ratna, Djoko M. Hartono, Setyo S. Mursidik, and Tri Edhi Budhi Soesilo. 2021. "Pollution Load Allocation on Water Pollution Control in the Citarum River." *Journal of Engineering & Technological Sciences* 53 (1).

Edgerton-Tarpley, Kathryn. 2016. North China famine, 1876–79. https://disasterhistory.org/north-china-famine-1876-79.

El-Sheekh, Mostafa, and Abd El-Fatah Abomohra. 2021. *Handbook of Algal Biofuels: Aspects of Cultivation, Conversion, and Biorefinery*, Elsevier.

EPA. "Basic Ozone Layer Science." Last modified October 7, 2021, accessed October 28. https://www.epa.gov/ozone-layer-protection/basic-ozone-layer-science.

EPA. "Flint Drinking Water Response." Last modified October 24, 2023, accessed October 26. https://www.epa.gov/flint.

EPA. "Superfund Site: Bunker Hill Mining & Metallurgical Complex Smelterville, ID." Accessed October 26. https://cumulis.epa.gov/supercpad/cursites/csitinfo.cfm?id=1000195&msspp=med.

EPA. "Basics of Climate Change," accessed October 22. https://www.epa.gov/climatechange-science/basics-climate-change.

EPA. 2022. "Climate Change Indicators: Wildfires." EPA, accessed October 24. https://www.epa.gov/climate-indicators/climate-change-indicators-wildfires.

EPA. 2021. "Climate Change Indicators: Tropical Cyclone Activity," Last Modified July 21, 2023, accessed October 24. https://www.epa.gov/climate-indicators/climate-change-indicators-tropical-cyclone-activity.

EPA Superfund Redevelopment Program. "Superfund Sites in Reuse in New Jersey." EPA, accessed November 4. https://www.epa.gov/superfund-redevelopment/superfund-sites-reuse-new-jersey.

Erdman, Jonathan and Wesner Childs, January 2023. "Hurricane Ian: Lessons Learned One Year Later." The Weather Channel, accessed October 23. https://weather.com/safety/hurricane/news/2023-09-21-florida-hurricane-ian-lessons-learned.

European Photovoltaic Industry Association. 2011. Solar Voltaics: Competing in the Energy Sector.

Farman, Joseph C., Brian G. Gardiner, and Jonathan D. Shanklin. 1985. "Large losses of total ozone in Antarctica reveal seasonal ClO x/NO x interaction." *Nature* 315 (6016):207–10.

Fernandez, Lucia. 2023. Geothermal energy generation worldwide in 2021, by country. Accessed November 11, 2023. https://www.statista.com/statistics/514488/geothermal-generation-worldwide-by-key-country/.

Fitzpatrick, T. B. 1976. "Halocarbons: Environmental effects of chlorofluoromethane release. Committee on impacts of stratospheric change. Assembly of Mathematical and Physical Sciences, National Research Council." Washington, DC: National Academy of Sciences:123.

Galindo, Andrea. 2022. What is Nuclear Energy? The Science of Nuclear Power. Accessed November 11, 2023. https://www.iaea.org/newscenter/news/what-is-nuclear-energy-the-science-of-nuclear-power.

Galt, Jerry A., William J. Lehr, and Debra L. Payton. 1991. "Fate and transport of the Exxon Valdez oil spill. Part 4." *Environmental Science & Technology* 25 (2):202–9.

Gantz, Carroll. 2015. *Refrigeration: A history*, McFarland.

Gates, Alexander E., and Robert P. Blauvelt. 2011. *Encyclopedia of pollution*, Facts on File.

Gates, Alexander E., and David Ritchie. 2006. *Encyclopedia of earthquakes and volcanoes*, Infobase Publishing.

Gates, Alexander. 2023. *Polluted Earth: The Science of the Earth's Environment*, John Wiley & Sons.

Gates, Alexander. 2023. *Polluted Earth: The Science of the Earth's Environment*, John Wiley & Sons.

Gliwa, Jana, Michael Wiedenbeck, Martin Schobben, Clemenz V. Ullmann, Wolfgang Kiessling, Abbas Ghaderi, Ulrich Struck, and Dieter Korn. 2022. "Gradual warming prior to the end-Permian mass extinction." *Palaeontology* 65 (5):e12621.

Gore, Al. 2006. *An Inconvenient Truth: The Planetary Emergency of Global Warming and What We Can Do about It*, Rodale.

Graham, F. Jr. 1978. "Rachel Carson." EPA, accessed October 26. https://www.epa.gov/archive/epa/aboutepa/rachel-carson.html.

Hartman, L. 2018. Wind Turbines in Extreme Weather: Solutions for Hurricane Resiliency. Accessed November 5, 2023.

Havlíček, Filip, and Miroslav Morcinek. 2016. "Waste and pollution in the ancient Roman Empire." *Journal of Landscape Ecology* 9 (3):33–49.

Hill, Jason. 2022. "The sobering truth about corn ethanol." *Proceedings of the National Academy of Sciences* 119 (11):e2200997119.

Hines, N. William. 2013. "History of the 1972 Clean Water Act: The story behind how the 1972 act became the capstone on a decade of extraordinary environmental reform." *George Washington Journal of Energy and Environmental Law* 4:80.

Hoesly, Rachel M., Steven J. Smith, Leyang Feng, Zbigniew Klimont, Greet Janssens-Maenhout, Tyler Pitkanen, Jonathan J. Seibert, Linh Vu, Robert J. Andres, and Ryan M. Bolt. 2018. "Historical (1750–2014) anthropogenic emissions of reactive gases and aerosols from the Community Emissions Data System (CEDS)." *Geoscientific Model Development* 11 (1):369–408.

Horn, Stuart A., and Captain Phillip Neal. 1981. "The *Atlantic Empress* sinking—a large spill without environmental disaster." International Oil Spill Conference.

Howard, Jenny. 2019. Dead zones, explained. *National Geographic*. Accessed November 11, 2023.

Idaho Department of Environmental Quality (IDEQ). "Bunker Hill Superfund Site." Accessed October 28. https://www.deq.idaho.gov/waste-management-and-remediation/mining-in-idaho/bunker-hill-superfund-site/.

International Tanker Owners Pollution Federation. 2023. "Oil Tanker Spill Statistics 2022." ITOPF Ltd., accessed October 31. https://www.itopf.org/knowledge-resources/data-statistics/statistics/.

Irving, P. M. 1988. "Overview of the U.S. National Acid Precipitation Assessment Program." Dordrecht.

Jacobs, Elizabeth T., Jefferey L. Burgess, and Mark B. Abbott. 2018. "The Donora smog revisited: 70 years after the event that inspired the clean air act." *American Journal of Public Health* 108 (S2):S85–S88.

Janousek, William M., Margaret R. Douglas, Syd Cannings, Marion A. Clément, Casey M. Delphia, Jeffrey G. Everett, Richard G. Hatfield, Douglas A. Keinath, Jonathan B. Uhuad Koch, and Lindsie M. McCabe. 2023. "Recent and future declines of a historically widespread pollinator linked to climate, land cover, and pesticides." *Proceedings of the National Academy of Sciences* 120 (5):e2211223120.

Joel, L. 2018. Ancient Romans polluted their lakes just like we do today. *Eos* 99. Accessed November 4, 2023. doi:10.1029/2018EO110747.

Johnsen, Reid, Jacob LaRiviere, and Hendrik Wolff. 2019. "Fracking, coal, and air quality." *Journal of the Association of Environmental and Resource Economists* 6 (5):1001–37.

Juuti, Petri S., Tapio Katko, and Heikkis Vuorinen. 2007. *Environmental History of Water*, IWA Publishing.

Katoh, T., T. Konno, I. Koyama, H. Tsurata, and H. Makino. 1990. "Acidic Precipitation in Japan." In *Acidic Precipitation: International Overview and Assessment*, edited by A. H. M. Bresser and W. Salomons, 41–105. New York, NY: Springer New York.

Kelley Blue Book. 2022. New-Vehicle Prices Set a Record in June, According to Kelley Blue Book, as Luxury Share Hits New High.

Kerstens, Sjoerd. 2013. Downstream impacts of water pollution in the upper Citarum River, West Java, Indonesia: Economic assessment of interventions to improve water quality. The World Bank.

Kim, Ki-Hyun, Ehsanul Kabir, and Shamin Kabir. 2015. "A review on the human health impact of airborne particulate matter." *Environment International* 74:136–143.

Körner, Sabine, Jan E. Vermaat, and Siemen Veenstra. 2003. "The capacity of duckweed to treat wastewater: Ecological considerations for a sound design." *Journal of Environmental Quality* 32 (5):1583–90.

Lark, Tyler J., Nathan P. Hendricks, Aaron Smith, Nicholas Pates, Seth A. Spawn-Lee, Matthew Bougie, Eric G. Booth, Christopher J. Kucharik, and Holly K. Gibbs. 2022. "Environmental outcomes of the US renewable fuel standard." *Proceedings of the National Academy of Sciences* 119 (9):e2101084119.

Leahy, P. Patrick, Joseph S. Rosenshein, and Debra S Knopman. 1990. *Implementation Plan for the National Water-quality Assessment Program*. Vol. 90: Department of the Interior, U.S. Geological Survey.

Lewis, J. 1985. "Lead Poisoning: A Historical Perspective." *EPA Journal*.

Liang, Youye. 2010. "A long lasting and extensive drought event over China in 1876–1878." *Advances in Climate Change Research* 1 (2):91–99.

Lindsey, Rebecca. 2023. "Climate Change: Atmospheric Carbon Dioxide." Climate. gov, accessed October 23. https://www.climate.gov/news-features/understanding -climate/climate-change-atmospheric-carbon-dioxide.

Lindsey, Rebecca. 2022. "Climate Change: Global Sea Level." Climate.gov, accessed October 23. https://www.climate.gov/news-features/understanding-climate/climate -change-global-sea-level.

Lindsey, Rebecca, and LuAnn Dahlman. 2023. Climate Change: Global Temperature. Climate.gov, https://www.climate.gov/news-features/understanding-climate/ climate-change-global-temperature. Accessed November 14, 2023.

Lindsey, Rebecca, and DLuAnn Dahlman. 2023. "Climate Change: Ocean Heat Content." climate.gov, accessed October 23. https://preview.climate.gov/news -features/understanding-climate/climate-change-ocean-heat-content.

Lippmann, Morton. 1989. "Health effects of ozone a critical review." *Japca* 39 (5):672–95.

Little, Jane Braxton. 2009. "The Ogallala Aquifer: Saving a vital US water source." *Scientific American* 1.

Luo, Taotao. 2016. USDA: Energy Efficiency of Corn-Ethanol Production Has Improved Significantly. Accessed November 6, 2023.

MacDonald, A. E., Clack, C. T., Alexander, A., Dunbar, A., Wilczak, J. and Xie, Y. 2016. "Future cost-competitive electricity systems and their impact on US CO2 emissions." *Nature Climate Change* 6 (5):526–31. doi: 10.1038/nclimate2921.

Mambra, S. 2022. The Complete Story of the Exxon Valdez Oil Spill. *Maritime History* 2023 (October 31).

Mann, Michael E. 2002. "Little ice age." *Encyclopedia of Global Environmental Change* 1 (504):e509.

Masmitjà, Gerard, Eloi Ros, Rosa Almache-Hernández, Benjamín Pusay, Isidro Martín, Cristóbal Voz, Edgardo Saucedo, Joaquim Puigdollers, and Pablo Ortega. 2022. "Interdigitated back-contacted crystalline silicon solar cells fully manufactured with atomic layer deposited selective contacts." *Solar Energy Materials and Solar Cells* 240:111731.

Marshall, Stephen. 2004. "The Meadowlands before the commission: Three centuries of human use and alteration of the Newark and Hackensack Meadows." *Urban Habitats* 2 (1):4–27.

Micklin, Philip. 2007. "The Aral Sea disaster." *Annual Review of Earth and Planetary Sciences* 35:47–72.

Mims, Christopher. 2008. "One Hot Island: Iceland's Renewable Geothermal Power." *Scientific American*, October 20, 2008.

Mintz, Joel A. 2011. "EPA enforcement of CERCLA: Historical overview and recent trends." *Southwestern Law Review* 41:645.

Molina, M. and Rowland, S. 1974. "Stratospheric sink for chlorofluorocarbons: Chlorine atom-catalysed destruction of ozone." *Nature* 249:810–12.

Morgan, J. D. 2011. "The Oil Pollution Act of 1990." *Fordham Environmental Law Review* 6 (1):1–27.

Musial, Walter, Paul Spitsen, Philipp Beiter, Patrick Duffy, Melinda Marquis, Rob Hammond, and Matt Shields. 2023. Offshore Wind Market Report: 2022 Edition. Washington, DC.

Narsilio, Guillermo Andres, and Lu Aye. 2018. "Shallow geothermal energy: An emerging technology." *Low Carbon Energy Supply: Trends, Technology, Management*:387–411.

NASA. 2022. "Global Temperature." accessed October 23. https://climate.nasa.gov/vital-signs/global-temperature/.

NASA. "Earth's Moon." accessed October 22. https://moon.nasa.gov/inside-and-out/dynamic-moon/weather-on-the-moon/.

National Research Council. 1976. *Halocarbons: Effects on Stratospheric Ozone*, National Academy of Sciences.

Needleman, Herbert L. 1999. "History of lead poisoning in the world." International conference on lead poisoning prevention and treatment, Bangalore.

Nelson, Gaylord. 1980. "Earth day '70: What it meant." *EPA Journal* 6 (4):6–38.

Nemery, Benoit, Peter H. M. Hoet, and Abderrahim Nemmar. 2001. "The Meuse Valley fog of 1930: An air pollution disaster." *The Lancet* 357 (9257):704–8.

Nikfar, S., and N. Rahmani. 2014. Valley of the Drums. In *Encyclopedia of Toxicology* edited by Philip Wexler. New York: Academic Press.

Nriagu, Jerome O. 1983. *Lead and Lead Poisoning in Antiquity*. New York and Chichester: J. Wiley.

NOAA. 2022. "Warm, dry October intensifies U.S. drought," accessed October 25. https://www.noaa.gov/news/warm-dry-october-intensifies-us-drought.

NOAA. 2011. "Extended Multivariate ENSO Index (MEI.ext)." Accessed October 24. https://psl.noaa.gov/enso/mei.ext/.

NOAA. 2023. "The Atmosphere," last modified July 28, 2023, accessed October 22. https://www.noaa.gov/jetstream/atmosphere.

Noakes, Tim J. 1995. "CFCs, their replacements, and the ozone layer." *Journal of Aerosol Medicine* 8 (s1):S-3-S-7.

NYS Department of Health. 1978. Love Canal—Public Health Time Bomb, edited by Health. Albany, NY: NYS Department of Health.

Oda, Takahiro, Jun'ya Takakura, Longlong Tang, Toshichika Iizumi, Norihiro Itsubo, Haruka Ohashi, Masashi Kiguchi, Naoko Kumano, Kiyoshi Takahashi, and Masahiro Tanoue. 2023. "Total economic costs of climate change at different discount rates for market and non-market values." *Environmental Research Letters* 18 (8):084026.

Owen, James. 2012. "World's largest dead zone suffocating sea." *National Geographic News* 5.

Patel, K. 2018. "Six trends to know about fire season in the western U.S." NASA Global Climate Change, accessed October 25. https://climate.nasa.gov/explore/ask-nasa-climate/2830/six-trends-to-know-about-fire-season-in-the-western-us/.

Patterson, C., Shirahata, H. and Ericson, J. 1987. "Lead in ancient human bones and its relevance to historical developments of social problems with lead." *Science of the Total Environment* 61:167–200. doi: 10.1016/0048-9697(87)90366-4.

Petrow, Richard. 1968. *In the Wake of Torrey Canyon*, D. McKay Company.

Pettis, Jeffery S., and Keith S. Delaplane. 2010. "Coordinated responses to honey bee decline in the USA." *Apidologie* 41 (3):256–63.

Rakhmat, Dikanaya Tarahita and Muhammad Zulfikar. 2018. Indonesia's Citarum: The World's Most Polluted River. *Asian Beat*. Accessed November 4, 2023.

Ramaswamy, Sonny. 2016. "Reversing pollinator decline is key to feeding the future." U.S. Department of Agriculture. Retrieved from https://www.usda.gov/media/blog/2016/06/24/reversing-pollinator-decline-key-feedingfuture.

Raynaud, Dominique, Jai Chowdhry Beeman, Jérome Chappellaz, F. Parrenin, and Jinhwa Shin. 2020. The long-term ice core record of CO2 and other greenhouse gases. https://scholar.google.com/scholar?hl=en&assdt=0%2C31&q=Raynaud%2C+Dominique%2C+Jai+Chowdhry+Beeman%2C+J%C3%A9rome+Chappellaz%2C+F.+Parrenin%2C+and+Jinhwa+Shin.+2020.+The+long-term+ice+core+record+of+CO2+and+other+greenhouse+gases&btnG=.

Reuben, Aaron, Maxwell Elliott, and Avshalom Caspi. 2020. "Implications of legacy lead for children's brain development." *Nature Medicine* 26 (1):23–25.

Revelle, Roger, and Hans E Suess. 1957. "Carbon dioxide exchange between atmosphere and ocean and the question of an increase of atmospheric CO2 during the past decades." *Tellus* 9 (1):18–27.

Rich, V. 1994. *The International Lead Trade*, Woodhead Publishing.

Ritchie, Hannah, and Max Roser. 2022. "Lead Pollution." *Our World in Data.*

Rodhouse, Thomas J., Rogelio M. Rodriguez, Katharine M. Banner, Patricia C. Ormsbee, Jenny Barnett, and Kathryn M. Irvine. 2019. "Evidence of region-wide bat population decline from long-term monitoring and Bayesian occupancy models with empirically informed priors." *Ecology and Evolution* 9 (19):11078–88.

Rodionova, Margarita V., Roshan Sharma Poudyal, Indira Tiwari, Roman A. Voloshin, Sergei K. Zharmukhamedov, Hong Gil Nam, Bolatkhan K. Zayadan, Barry D. Bruce, Harvey J. M. Hou, and Suleyman I. Allakhverdiev. 2017. "Biofuel production: Challenges and opportunities." *International Journal of Hydrogen Energy* 42 (12):8450–61.

Rogan, W. J., Bagniewska, A., and Damstra, T. 1980. "Pollutants in breast milk." *New England Journal of Medicine* 302:1450–53.

Rosenberg, Kenneth V., Adriaan M. Dokter, Peter J. Blancher, John R. Sauer, Adam C. Smith, Paul A. Smith, Jessica C. Stanton, Arvind Panjabi, Laura Helft, and Michael Parr. 2019. "Decline of the North American avifauna." *Science* 366 (6461):120–24.

Roser, Max and Ritchie, Hannah 2023. How has world population growth changed over time? *Our World in Data.*

Rothschild, Rachel Emma. 2019. *Poisonous Skies: Acid Rain and the Globalization of Pollution*, University of Chicago Press.

Russo, Danny. 2023. EV battery replacement cost. *Consumer Affairs*. Accessed November 6, 2023.

Sánchez-Bayo, Francisco, and Kris A. G. Wyckhuys. 2019. "Worldwide decline of the entomofauna: A review of its drivers." *Biological Conservation* 232:8–27.

Savage, E. P., T. J. Keefe, J. D. Tessari, H. W. Wheeler, F. M. Applehans, E. A. Goes, and S. A. Ford. 1981. "National study of chlorinated hydrocarbon insecticide residues in human milk, USA: I. Geographic distribution of dieldrin, heptachlor, heptachlor epoxide, chlordane, oxychlordane, and mirex." *American Journal of Epidemiology* 113 (4):413–22.

Scheader, Edward C. 1991. "The New York City water supply: Past, present and future." *Civil Engineering Practice* 6 (2):7–20.

Schippers, Peter, Ralph Buij, Alex Schotman, Jana Verboom, Henk van der Jeugd, and Eelke Jongejans. 2020. "Mortality limits used in wind energy impact assessment underestimate impacts of wind farms on bird populations." *Ecology and Evolution* 10 (13):6274–87.

Schmalensee, Richard, and Robert N. Stavins. 2019. "Policy evolution under the clean air act." *Journal of Economic Perspectives* 33 (4):27–50.

Scully, M. J., Norris, G. A., Alarcon Falconi, T. M. and MacIntosh, D. L. 2021. "Carbon intensity of corn ethanol in the United States: state of the science." *Environmental Research Letters* 16. doi: 10.1088/1748-9326/abde08.

Seibold, Sebastian, Martin M. Gossner, Nadja K. Simons, Nico Blüthgen, Jörg Müller, Didem Ambarlı, Christian Ammer, Jürgen Bauhus, Markus Fischer, and Jan C. Habel. 2019. "Arthropod decline in grasslands and forests is associated with landscape-level drivers." *Nature* 574 (7780):671–74.

Selby, Karen. "Mesothelioma Death and Mortality Rate." Asbestos.com, last modified September 29, 2023, accessed October 29. https://www.asbestos.com/mesothelioma/death-rate/.

Shara, S., S. S. Moersidik, and T. E. B. Soesilo. 2021. "Potential health risks of heavy metals pollution in the downstream of Citarum River." IOP Conference Series: Earth and Environmental Science.

Sharma, R. 2019. "Effect of obliquity of incident light on the performance of silicon solar cells." *Heliyon* 5 (7).

Sheehan, John, Vince Camobreco, James Duffield, Michael Graboski, Housein Shapouri. 1998. An Overview of Biodiesel and Petroleum Diesel Life Cycles. U.S. Department of Agriculture and U.S. Department of Energy.

Shine, I. 2022. The world needs 2 billion electric vehicles to get to net zero. But is there enough lithium to make all the batteries? Accessed November 5, 2023.

Sholeh, Muhammad, Pranoto Pranoto, Sri Budiastuti, and Sutarno Sutarno. 2018. "Analysis of Citarum River pollution indicator using chemical, physical, and bacteriological methods." AIP Conference Proceedings.

Skene, Jennifer, and Vinyard, Shelley. February 20, 2019. The Issue with Tissue: How the U.S. Is Flushing Forests Away, Natural Resources Defense Council(NRDC), https://www.nrdc.org/bio/jennifer-skene/issue-tissue-how-us-flushing-forests-away.

Slaper, Harry, Guus J. M. Velders, John S. Daniel, Frank R. de Gruijl, and Jan C. van der Leun. 1996. "Estimates of ozone depletion and skin cancer incidence to examine the Vienna Convention achievements." *Nature* 384 (6606):256–58.

Smith, Daniel. 1999. "Worldwide trends in DDT levels in human breast milk." *International Journal of Epidemiology* 28 (2):179–88.

Smith, Kristine M., Elizabeth H. Loh, Melinda K. Rostal, Carlos M. Zambrana-Torrelio, Luciana Mendiola, and Peter Daszak. 2013. "Pathogens, pests, and economics: drivers of honey bee colony declines and losses." *EcoHealth* 10:434–45.

Soll, David. 2013. *Empire of Water: An Environmental and Political History of the New York City Water Supply*, Cornell University Press.

Solly, Ray. 2022. "The Development of Crude Oil Tankers: A Historical Miscellany." *The Development of Crude Oil Tankers*:1–192.

Song, Haijun, David B. Kemp, Li Tian, Daoliang Chu, Huyue Song, and Xu Dai. 2021. "Thresholds of temperature change for mass extinctions." *Nature Communications* 12 (1):4694.

Squyres, S. W. "Venus." Encyclopedia Britannica, last modified October 19, 2023, accessed October 22, 2023. https://www.britannica.com/place/Venus-planet.

Stets, Edward G., Lori A. Sprague, Gretchen P. Oelsner, Hank M. Johnson, Jennifer C. Murphy, Karen Ryberg, Aldo V. Vecchia, Robert E. Zuellig, James A. Falcone, and Melissa L. Riskin. 2020. "Landscape drivers of dynamic change in water quality of US rivers." *Environmental Science & Technology* 54 (7):4336–43.

Struck, D. 2009. Twenty Years Later, Impacts of the Exxon Valdez Linger. *Yale Environment 360*.

Thordarson, Thorvaldur, and Stephen Self. 2003. "Atmospheric and environmental effects of the 1783–1784 Laki eruption: A review and reassessment." *Journal of Geophysical Research: Atmospheres* 108 (D1):AAC 7-1-AAC 7–29.

Tilton, George. 1998. "Clair Cameron Patterson " In *Biographical Memoirs*, 266–287. Washington, DC: The National Academies Press.

Tormey, Szilvia Doczi and Jennifer Chen. 2019. U.S. regulatory innovation to boost power system flexibility and prepare for ramp up of wind and solar. IEA, https://www.iea.org/commentaries/us-regulatory-innovation-to-boost-power-system-flexibility-and-prepare-for-ramp-up-of-wind-and-solar. Accessed November 7, 2023.Tsai, W. T. 2014. Chlorofluorocarbons. In *Encyclopedia of Toxicology* edited by Philip Wexler, Academic Press.

Turgeon, Andrew and Elizabeth Morse. Geothermal Energy. Accessed November 11, 2023. https://education.nationalgeographic.org/resource/geothermal-energy/.

Turner, R. Eugene. 2021. "Declining bacteria, lead, and sulphate, and rising pH and oxygen in the lower Mississippi River." *Ambio* 50 (9):1731–1738.

Turusov, Vladimir, Valery Rakitsky, and Lorenzo Tomatis. 2002. "Dichlorodiphenyltrichloroethane (DDT): Ubiquity, persistence, and risks." *Environmental Health Perspectives* 110 (2):125–28.

USGS Water Resources Mission Area. 2019. Water Quality in the Nation's Streams and Rivers: Current Conditions and Long-Term Trends. Accessed November 4, 2023.

Umar, Sheikh Ahmad, and Sheikh Abdullah Tasduq. 2022. "Ozone layer depletion and emerging public health concerns-an update on epidemiological perspective of the ambivalent effects of ultraviolet radiation exposure." *Frontiers in Oncology* 12:866733.

University of Buffalo Library. Love Canal: Timeline and Photos. Buffalo, NY.

USDOE, Alternative Fuels Data Center. 2023. Electric Vehicle Benefits and Considerations. Accessed November 6, 2023.

U.S. Department of State. The Montreal Protocol on Substances That Deplete the Ozone Layer.

U.S. Energy Information Administration. "Oil and Petroleum Products Explained." U.S. Department of Energy, last modified June 12, 2023, accessed October 31. https://www.eia.gov/energyexplained/oil-and-petroleum-products/.

U.S. Fish and Wildlife Service. "Rachel Carson (1907–1964) Author of the Modern Environmental Movement." Accessed October 26. https://www.fws.gov/staff-profile/rachel-carson-1907-1964-author-modern-environmental-movement.

U.S. Fish and Wildlife Service. 2021. "Fact sheet: Bald Eagle Haliaeetus leucocephalus." Accessed October 26. https://www.fws.gov/sites/default/files/documents/bald-eagle-fact-sheet.pdf.

U.S. Fish and Wildlife Service. 1999. "Peregrine Falcon (Falco peregrinus)." Accessed October 26. http://npshistory.com/brochures/nwr/wildlife-fact-sheets/peregrine-falcon-1999.pdf.

U.S. Fish and Wildlife Service. 2009. "Fact sheet: Brown Pelican Pelecanus occidentalis." Accessed October 26. https://www.fws.gov/sites/default/files/documents/brown_pelicanfactsheet09.pdf.

Vallero, Daniel A. 2014. *Fundamentals of Air Pollution*, Academic Press.

Vaughn, A. 2017. Torrey Canyon disaster—the UK's worst-ever oil spill 50 years on. Accessed October 31, 2023.

Wagner, David L., Eliza M. Grames, Matthew L. Forister, May R. Berenbaum, and David Stopak. 2021. "Insect decline in the Anthropocene: Death by a thousand cuts." *Proceedings of the National Academy of Sciences* 118 (2):e2023989118.

Waldron, Harry A. 1973. "Lead poisoning in the ancient world." *Medical History* 17 (4):391–99.

Waseem, Hafiza Hafza, Asma El Zerey-Belaskri, Farwa Nadeem, and Iqra Yaqoob. 2016. "The downside of biodiesel fuel–a review." *International Journal of Chemical and Biochemical Sciences* 9:97–106.

Wegmann, Edward. 1896. *The Water-supply of the City of New York. 1658–1895*, J. Wiley & Sons.

Wells, P. G. 2017. "The iconic Torrey Canyon oil spill of 1967: Marking its legacy." *Marine Pollution Bulletin* 115 (1/2):1–2.

Williamson, Marcus. 2012. Professor Sherwood Rowland Scientist who helped establish CFCs' harmful effects. *Independent*. Accessed October 28.

World Meteorological Organization, WMO. 2018. Scientific Assessment of Ozone Depletion: 2018. In *Global Ozone Research and Monitoring Project* Geneva, Switzerland.

World Nuclear Association. "Supply of Uranium." World Nuclear Association, last modified August 2023, accessed November 11. https://world-nuclear.org/information-library/nuclear-fuel-cycle/uranium-resources/supply-of-uranium.aspx.

Wu, Yuyang, Daoliang Chu, Jinnan Tong, Haijun Song, Jacopo Dal Corso, Paul B. Wignall, Huyue Song, Yong Du, and Ying Cui. 2021. "Six-fold increase of atmospheric p CO_2 during the Permian–Triassic mass extinction." *Nature Communications* 12 (1):2137.

Zattara, Eduardo E., and Marcelo A. Aizen. 2021. "Worldwide occurrence records suggest a global decline in bee species richness." *One Earth* 4 (1):114–23.

Index

About the Author

Alexander Gates is a distinguished service professor of earth and environmental sciences at Rutgers University in Newark, New Jersey. Dr. Gates holds a PhD in geology from Virginia Tech. He has published nine books and eighty-one professional papers and has edited eleven professional volumes. His work in geology, education, and diversity improvement in STEM has been recognized with twenty-six professional awards.